PENGUIN LIFE

Why We Eat (Too Much)

'I would recommend [it] to anyone trying to lose weight. I would also recommend [it] to those who struggle to be kind to people with obesity' *The Times*

'A compelling look at the science of appetite and metabolism' *Vogue*

'A refreshing antidote to junk public health diet advice. The current epidemic of obesity and diet-related disease is a legacy of catastrophic public health advice based on flimsy science. Dr Jenkinson takes down this failed paradigm and replaces it with sensible, workable advice that reflects the sanity of new-wave nutrition thinking. Articulate, clear, a joy to read, this is a book that really needed to be written' Joanna Blythman, author of *Swallow This*

'Will engross anyone who has ever struggled with dieting, but you don't have to want to lose weight to read it . . . If you do this then 2020 might be the year that you learn to understand your body' *Sunday Telegraph*

'Dr Jenkinson wants solutions because he knows how recent and overwhelming the global weight problem is . . . His message is for individuals wanting help, not industry or government who may block their ears. It's one of enjoyment, not deprivation, summed up by "eat more, rest more"' Jenni Russell, *The Times*

'Distilling what he has learned over decades of practice [. . .] in his book, Mr Jenkinson says a far better approach is to ditch the quick-fix solution in favour of an old-fashioned approach' *Mail on Sunday*

'Fascinating science . . . One of Britain's top weight-loss experts' ITV

'Cool, clear and highly persuasive . . . A radical approach to weight loss' *Sunday Times*

Dr Andrew Jenkinson is a consultant in bariatric (weight loss) and general surgery at University College London Hospital. He has travelled around the globe, presenting his research and teaching bariatric surgery. He runs a regular weight-loss clinic in Dubai. He specializes in weight loss and diet and has been performing bariatric surgery for fifteen years – seeing, treating and most importantly, talking to over three thousand patients in this time.

WHY WE EAT
(TOO MUCH)

The New Science of Appetite

DR ANDREW JENKINSON

PENGUIN LIFE

AN IMPRINT OF

PENGUIN BOOKS

PENGUIN LIFE

UK | USA | Canada | Ireland | Australia
India | New Zealand | South Africa

Penguin Life is part of the Penguin Random House group of companies
whose addresses can be found at global.penguinrandomhouse.com.

First published 2020
This edition published with new material 2021
001

This book is based on my own experience as a consultant surgeon with a special
interest in advanced laparoscopic, or keyhole, surgery. In order to protect the
privacy and confidence of patients, I have, of course, not used any real names and
I have changed all physical descriptions and other features to ensure that no one
is identifiable. I have, in some cases, also changed genders and racial origins.
This is because this is not a book about the individuals I have described,
but about what we can learn from them.

Printed and bound in Great Britain by Clays Ltd, Elcograf S.p.A.

The authorized representative in the EEA is Penguin Random House Ireland,
Morrison Chambers, 32 Nassau Street, Dublin D02 YH68

A CIP catalogue record for this book is available from the British Library

ISBN: 978–0–241–40053–1

www.greenpenguin.co.uk

CONTENTS

PART THREE: BLUEPRINT FOR A HEALTHIER WEIGHT
The Secret to Lasting Weight Loss

LIST OF ILLUSTRATIONS

INTRODUCTION

Bariatric Surgery Outpatients Clinic, London, December 2012

Clinic K is where people come to discuss having their stomach removed.

The office, occupying a whole corner of University College Hospital's first floor, has a backdrop of London through large floor-to-ceiling windows. Looking out onto the red buses and black cabs of Euston Road, I remember recognizing one of my patients slowly making her way over to the hospital's main entrance, sheltering her large body under a flapping umbrella in the storm, vainly trying to keep dry for her appointment. I felt sad for her.

Minutes later she entered, trepidation and despair etched on her face. She had finally given in, held up the white flag, surrendered in her battle with her weight – she had lost the diet wars. She wanted me to remove most of her stomach. She eased herself into our over-sized clinic chair and tearfully recounted her years of dietary failure. And as she talked I listened and learned.

Why We Eat (Too Much) was inspired by patients just like this lady – normal people who had suffered with their weight for years; people who came to me looking for treatment.

My patients encouraged me to write this book. I had listened to them over the years and what they said did not fit with my own understanding of obesity. I wanted to close this gap between what scientists, doctors and dieticians told us about it, and how to deal with it, and what obese people actually experienced – because the two stories did not fit together. Someone had got it wrong.

If, as the scientists told us, it was simple to lose weight by dieting and exercise, and if the benefits of that weight loss were so great in terms of happiness, confidence, health and finances, then why could people not achieve it? Over the next five years I became intrigued by this question: why something seemingly so simple could in practice prove to be so difficult. Why can't people sustain weight loss? How could weight loss through dieting be so difficult that people resorted to such extreme measures as stomach-removal (or bypassing) surgery?

University College London Hospital (UCLH) has a fantastic metabolic research unit, run by my colleague Professor Rachel Batterham. Her cutting-edge research gave me a head-start in understanding how our appetite is controlled by strong hormones (originating in the stomach and intestines) that have a profound effect on what we eat and how much we eat. Appetite did not seem to be under a great deal of conscious control; instead, it was governed by these newly discovered hormones.

My studies took me from appetite on to metabolism. How is the amount of energy that we burn controlled? More hormones seemed to be involved. But, curiously, much of the ground-breaking research that explained our metabolism was being ignored by mainstream medicine. Why was this?

If our appetite and our metabolism are being controlled by powerful hormones, then this would explain why it is so difficult for my patients to lose weight using simple willpower. The hormonal triggers that drive our eating and resting behaviour seem to be mainly influenced by our changing environment.

In this book, I will use the newly emerging scientific understanding of metabolism and appetite and combine this knowledge with what obese people have been trying to tell us for years. I will explain why most of the things that you have been told about obesity are myths, based on poor research and vested interests. I will explain:

- Why it is so difficult to lose weight if you use the current advice from medical and nutritional experts
- How some of this dietary advice can be counter-productive and make weight loss even more difficult
- The best strategies for long-term weight loss and health, whether you want to lose 5lb or 5 stone
- Why many very obese people have the feeling that they are trapped and cannot escape, no matter how hard they try.

Once you have read this book you should have a better understanding of why medical professionals have been failing in their advice on weight loss for so many years and, more importantly, you will be able to use this knowledge to improve your own health and well-being. By the end of this book I hope that you will feel a sense of relief that, finally, you have not only an explanation, but also a solution. I will avoid excessive medical jargon (and explain any terms that need to be used) and present my ideas in an accessible (sometimes light-hearted) way to keep you reading.

But first, some background. I am a surgeon at University College Hospital in London. My job is to treat people who cannot lose weight by dieting and have reached the end of the road. They have accepted that for them it is impossible to lose weight and keep that weight off. They know that unless something drastic is done they will spend their lives feeling trapped beneath layers of fat, slowly getting sicker and more frustrated and unhappy. In the last fifteen years I must have interviewed over 2,000 people in this situation.

The solution my patients seek is surgery. Not surgery like lipo-suction to suck out the fat, but an operation that will change their stomach and intestines to make it easier for them to lose weight: bariatric surgery. You may have heard about this type of surgery in the media. A popular bariatric operation is the 'gastric band'. This entails having an adjustable band (made of a type of plastic) placed around the upper part of the stomach. The band works by stopping you eating very fast – making you feel full (and sometimes uncomfortable) after a very small meal. The gastric band has now been overtaken in popularity by two other procedures: one where the stomach is completely bypassed (so food doesn't ever see the stomach), and another where three quarters of the stomach is completely removed, leaving what remains in the shape and size of a narrow tube. This is called the sleeve gastrectomy (more on this in chapter 6).

My first weight-loss operation was a gastric bypass in 2004, using laparoscopic, or keyhole, surgery. This is quite a difficult procedure to perform. I had been trained well, but when the morning of the operation came, and I saw my patient, I was anxious for him. He was high risk: a very large 210kg (33-stone) young orthodox Jewish chef called Jac.

The surgery went well. It had taken two and a half hours, although it hadn't felt that long. Once you are performing a procedure you concentrate so hard it is as if you have been sent to a different world.

When you begin, you do not usually feel nervous about all the responsibility because you know that you should be able to deal with most problems if they arise. Performing an operation, especially if you have become familiar with it, can almost be a meditative, deeply relaxing experience.

Jac made a great recovery, and because keyhole surgery means that there are no large cuts in the abdomen – only small nicks – the pain afterwards is minimal. Happily, he walked out of the hospital pain-free soon after surgery.

Many of my fellow doctors think that bariatric procedures are unnecessary and mutilating. They think, or say, 'Why can't your patients just lose weight on a diet and have a little bit more will-power?' And it is not just doctors who think this. Many politicians and journalists, people who wield real power, would also argue that this type of surgery should not really be necessary, or available. My belief is that they are wrong. This book sets out to explain our fundamental misunderstanding of the causes of and treatments for obesity. And it is because of this flawed thinking by many experts and advisers that the obesity crisis has become worse and anyone who suffers from it gets more frustrated. If we, as a society, understood obesity and came together to tackle it, then we would not need my services, or the services of any weight-loss surgeon.

After my first successful operation in 2004, I started to do more and more of this type of bariatric surgery: gastric bypass, gastric bands and sleeve gastrectomies. As I became more proficient at them, the Homerton University Hospital, where I first started as a consultant, grew to be the busiest weight-loss surgery centre in London. With experience, the time I required for an operation was reduced to one hour and most patients needed to stay in the hospital for only one night and then needed just a week off work.

As the months and years passed, my outpatient clinic became increasingly swamped with patients suffering with different degrees of obesity. I spoke to many hundreds of patients about their views of the condition and what they had experienced first-hand. Then I had a revelation: they all seemed to be saying the same things to me repeatedly. There was no collusion between patients – they did not know what others might have been saying. Their views and their experiences of obesity went against the conventional view of doctors, dieticians and other health professionals. As they vocalized their thoughts, I started to listen and think.

I recalled the teaching of David Maclean, an immaculately dressed surgeon I had worked with at the Royal London Hospital who, at the age of sixty-eight, had carried on working past retirement age because they could not find an adequate replacement for him. He would look me in the eyes and say, 'Always listen *closely* to what your patients are telling you.' This advice stayed with me – I listened. These were some of the typical things that I heard, time and time again:

- 'I can lose weight, doctor, but I can never keep it off'
- 'I think that I have a slow metabolism compared to other people I live with'
- 'I think obesity is in my genes'

or

- 'Diets don't work for me, I have tried them all and I end up gaining more weight than when I started the diet'
- 'I only have to look at a cream cake and I get fat!'
- 'I can't control my hunger, I feel weak if I don't eat.'

When I first started doing these clinics I relied on my limited training in obesity from medical school. I had become very good at doing the operations to treat the condition but, as with many doctors confronted with a patient suffering with obesity, I had poor empathy – I didn't *really* appreciate what they were experiencing. I understood the simple principle of energy balance – if you take in more energy in the form of (food) calories than you burn off (through exercise), then you will store that extra energy inside your body as fat. Therefore, in my mind, it was very simple to lose weight. You merely had to eat less and exercise more – that's how us medics understood it, but it didn't seem that simple to my patients.

What also struck me in those first few years of treating obesity was the transformation of my patients after the surgery. Their lives had been turned around. This condition, obesity, that they had been fighting all their lives, was no longer present. Many said they were back to their former selves – their pre-obese selves. The problem that they had been trying to deal with for years and years, with diet after diet, and disappointment after disappointment, was now gone. They had been released from their obesity trap.

Realizing that each of my patients was telling me virtually the same story before surgery, and that they had become different people after surgery, I started to wonder whether what they, the patients, were saying was correct, and what we, the doctors, were saying was wrong – whether our conventional understanding of obesity was flawed. Was this a condition that arose in patients without them having any control over its development? In other words, was it more of a disease than a lifestyle choice? I wanted answers to these questions.

The tabloid journalists, the doctors, the policy-makers, the public and the politicians were pointing their fingers at my patients and saying, 'This is your problem, you made it and if you had enough willpower you could solve it.' But my patients were giving me a different message: 'I will do anything, but I am trapped.' So I thought that I should try and establish the truth. What if my patients were right and the medical establishment was wrong? I went back to the books and I studied and researched the whole area of metabolism, weight regulation and appetite. I wanted to square what I had heard and seen from my years of speaking to and treating obese patients with what was in the medical research literature. I embarked on a further journey into the depths of metabolic research, into the genetics and epigenetics of obesity, how anthropology, geography and economics affected our foods, and how scientists and lobbyists influenced our understanding.

Once I had done my research I had my answer. Patients liked my way of explaining to them exactly why they were trapped in this condition. How weight is not under conscious control and therefore cannot be manipulated down by dieting. How you should encourage the body to want to be lighter by changing the day-to-day signals it receives. This is the basis of *Why We Eat (Too Much)*.

I hope that this book will be read by anyone who is interested in controlling their weight, but is tired of dieting. I hope that people who want to fully understand obesity and weight regulation will pick up this book – anyone who has a friend or relative who is struggling with obesity and cannot control it. Finally, I hope that the people in power – the politicians, journalists and (dare I say it?) even doctors – will study this book. It will change your perception of obesity and, maybe, help future generations to avoid the suffering it brings.

PART ONE

Lessons in Energy

How Our Body Works to Control Weight

ONE

Metabology for Beginners

How Our Weight is Controlled

Humans talk, write, walk, and love using the same amount of
energy per second as a light bulb, a device that does nothing but
shine light and get hot. This amazing fact, far from denigrating
humans, is a testament to how efficient a human body is. But
more importantly, it is a testament to the wondrous complexity
of our bodies, which can do so much with so little.

Peter M. Hoffmann, *Life's Ratchet: How Molecular*
Machines Extract Order from Chaos (2012)

I distinctly remember the first class on my first day of medical school.
We were issued with starched white coats to cover up our student
sweaters and ripped jeans. The superintendent ushered us into a
bright neon-lit room, chilled like a walk-in fridge. Along the length of
the room were many evenly placed narrow tables, each with a cotton
sheet obscuring what lay below. We paired up, took a table each and
jokily grappled to get our latex gloves on. One hour later, if you could
have observed this group of eighteen-year-olds coming out of their
first lesson, you would have noticed several distinct differences from
when they walked in. Two of the group needed to be helped out of
the room – and did not consider a career in medicine again. The
rest of us were ashen-faced. The sheets over each table had hidden a
human cadaver. Each one was drained of blood, with a shaven head,
the bodies grey in colour and infused with acrid-smelling formalin
to preserve them. This was our first lesson: ANATOMY.

During our anatomy classes of that year, we dissected, and examined, all the different organs that run the body. We learned how each

individual part of the body worked to maintain health. The organ systems that we learned about included:

- Cardiology – how the heart and circulation work
- Pulmonology – how the lungs oxygenate our blood
- Gastroenterology – how we digest and absorb food
- Urology – how the kidneys maintain fluid balance in our body
- Endocrinology – how glands and hormones work.

These systems gave us the grounding to eventually understand the entire workings of the human body and a basis from which to go on and learn about the diseases that affected them. The classes were supposed to cover all the diseases that we would encounter in our future careers as doctors. However, there was one major omission – none of the organ systems that we learned about adequately explained obesity, the condition that would go on to affect a quarter of our patients throughout our careers, and would trigger unprecedented levels of diabetes, blood pressure and heart problems.

When we took our sharp scalpels in hand and dissected our cadaver, the first layers to be discarded were the skin and fat. These handfuls of human jelly were thrown into bins for later incineration. What we were unaware of at the time was that by getting rid of the fat we were rejecting an important part of the body. Where was the organ that controls our metabolism and appetite; that coordinates and stores our energy reserves? As we busily dissected a lung, a heart or a kidney, this vital organ was in the tissue bin – discarded and ignored.

Have medical schools now caught up? When I quiz my students on the training they currently receive in order to understand obesity, it remains similar to the curriculum of the 1980s, with only minor changes. Experts in obesity are therefore by definition self-taught, and because of this their views often differ from regular doctors, who still rely on the limited training they received in medical school.

In this book, we will go to my 'virtual' medical school to cover the subject that should be on the curriculum, but sadly remains ignored. So let's give the subject a brand new medical name: *metabology*, from the prefix *metabo-*, for 'metabolism', the chemical processes in cells related to energy, and the suffix *-logy*, meaning 'the study of'.

Metabology – the study of appetite and metabolism, of fat storage or fat loss; the study of the energy flows into and out of the body.

Metabology is simple – there are only two main rules that you need to remember to master it. You know one of these rules already – energy in (food) minus energy out (exercise) equals energy stored (usually fat). But the other rule is less widely understood. It states that our bodies try and maintain a healthy internal environment by a process called negative feedback. This is the way the body works to stop you losing (or gaining) weight too fast. Remember these rules and you will understand obesity and its causes and treatment better than most. You will have a superior understanding of obesity compared with most doctors, and if you have struggled with weight control in the past all those struggles will become much clearer.

Before we discuss the Two Rules of Metabology in more detail, let's first take a look at that organ that was thrown away into the incineration bin in the anatomy class – fat. Fat, or adipose tissue as it is known in medical language, is now recognized as one of our vital and life-preserving organs. An organ is defined as being part of a living thing, but separated from other parts, and having a specific function. The specific function of fat is energy regulation. As we will see, fat not only stores energy but also controls how much we use.

A Light, Insulating Energy Source

Fat is made up of individual cells called adipocytes. These cells play a critical role in the survival of any mammalian species – from seals to camels to humans. It has three major properties. First, it is light, compared to muscle or bone; therefore it is efficient to carry around. Second, it provides insulation against the cold and therefore prevents too much thermal or heat energy loss to the air, especially in cold climates. Handy if you are a seal enveloped in a thick layer of blubber, swimming around in ice-cold oceans, not so handy if you are a camel in the 40°C heat of the desert – unless of course you store all the fat in one big lump, or hump, and let the rest of the body breathe. Third, it can store large amounts of energy. Fat is an efficient, lightweight, insulating energy source.

Each fat cell has the unique ability to store energy for times when it may be needed. The more energy it stores, the more bloated it becomes and the more the fat cell *expands in size*. In the initial process of becoming fatter, you do not grow more fat cells. The number of cells stays the same, but each one becomes swollen with its stored energy, growing to six times its original volume. When there is no more room within the cells, the number of fat cells in the body increases – from an average of 40 billion to over 100 billion in some cases. Unfortunately, if you suck the fat cells out with liposuction (a common, short-term fix performed by plastic surgeons), more and more fat cells are produced to compensate.

Energy storage is the most important function of fat as an organ. It is critical to have a store of energy to survive in times of famine and food shortage. The brain needs a constant level of glucose (sugar) in the blood to function. When there is no food readily available, this is replenished constantly by our fat cells. In many types of mammals, including man, there does not need to be an actual famine for our fat stores to be called upon. During migrations, fights to defend territory, fights to obtain a mate, the act of mating, pregnancy and breast-feeding, the amount of energy taken in as food can be reduced even though the amount of energy needed increases. This is when the fat-storage function comes into play. An energy bank in the form of fat, like a fuel tank in a car, is critical to our survival and for our ability to reproduce and raise the next generation.

You might therefore think that there would be a major evolutionary advantage to having a large energy store. However, it isn't in your interest to be carrying an oil-tanker's worth of energy around as this is going to limit your ability to go about your normal survival activities like hunting or running away from hungry predators. So there must be a mechanism to control the size of these fat tanks: fat is very clever, and very efficient, at self-regulation.

Metabology Rule 1 – Energy Use and Storage

The first rule to remember is already in the curriculum for medical students. In most people's opinion, this rule is what defines obesity: it explains, simply and precisely, energy use and storage. But it is this rule that causes so much prejudice against people who struggle with weight control. It is grandly named 'The First Law of Thermodynamics' and is used by physicists to calculate the amount of energy stored in any given object – from a rock, to a plant, to an animal (including a human). Its basic premise is: the energy stored in an object equals the amount of energy taken in minus the amount of energy taken out.

If you want to simplify things, then just think of a human as a box. This box transforms chemical energy from food into heat, movement and thought. The rest is stored.

(Energy In) – (Energy Out) = Energy Stored

In humans, the 'Energy In' is what we eat – a combination of proteins, fats and carbohydrates. The 'Energy Out' part is just as important and is often misunderstood. Often people think that most of the energy they use up comes from how active they are in the daytime and whether they go to the gym or not. This is not the case. Most of the energy that we use does not involve any type of movement. If we were to lie in bed all day and all night we would still use up to 70 per cent of the energy that we normally do – through breathing, heartbeat, temperature control and all our cells' chemical reactions. The amount of energy that we use to perform these subconscious tasks is called our *basal metabolic rate* (or BMR). The concept that over two thirds of our daily energy expenditure is not within our conscious control is an important one to grasp when understanding our metabolism – and how we control our weight and why some people develop obesity.

What about the remaining 30 per cent of the energy that we normally use? This is made up of two parts:

1. *Passive energy expenditure* – the energy that we use to get on with our everyday lives. This can be anything from walking to work, doing the cleaning, moving around the office or doing a hobby. For most of us – those who don't go to the gym or have a manual job – this will make up almost all of the remaining 30 per cent of energy used.
2. *Active energy expenditure* – this is the amount of energy that we use up when we perform active exercise. For some this could be going to the gym or jogging. For others, such as builders in England, rickshaw drivers in India, or hunters in the African savannah, it could be part of their daily lives. For sedentary people, meaning most of us working in cities, active energy expenditure may just be running for a bus, or climbing a few flights of stairs, and accounts for just 2 or 3 per cent of our total daily energy used.

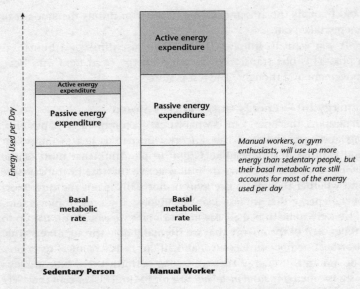

Figure 1.1 Energy used per day by sedentary people compared with active people

Manual workers, or gym enthusiasts, will use up more energy than sedentary people, but their basal metabolic rate still accounts for most of the energy used per day

FACT BOX

The sugar energy in the liver needs water to hold it in place; this makes it quite a heavy energy source (water is much heavier than fat). When you go on a very low-calorie diet, the liver's stores of energy are the first to be used up. As the sugar in the liver is used, so the water is flushed out and hey presto you think that you have lost a lot of weight in a few days – but it is mostly water and not fat. This is one of the main tricks that fad diets play on people: you think you are making real progress with initial weight loss but it is mainly fluid and the weight loss is transient.

The 'Energy Stored' part of the equation is simpler. Any excess energy is stored first in the liver (as a type of sugar) and then in fat cells (as fat). The liver can only hold a couple of days' worth of energy; it is generally full to capacity, so in practice excess energy is usually stored in fat. The energy in fat can help keep us going for about thirty days without food. Knowing this takes us on to the rule that is almost always overlooked when explaining obesity.

Metabology Rule 2 – Negative Feedback System

The second rule is called the *negative feedback system*. You may wonder, isn't that what I get from the boss when he catches me coming into work late? And yes, in a way you would be correct. Negative feedback describes the regulation of a system: it can be an office system or a mechanical system (like a machine) or a biological system (like that of a human). The system has a set way of working (like nine-to-five office hours) and if it senses the way of working deviates from the set rule, then it will automatically correct itself.

Negative feedback systems are simple. They just need a sensor connected to a switch which changes the system back to where it should be. In our office example, the boss is the sensor to your late arrival and his warning is the switch to change your future behaviour.

An example of this in a machine would be a household thermostat. It is designed to maintain a set temperature. It senses when the temperature in the house falls below this and switches on the central heating. When the temperature then exceeds the setting, it automatically turns the heating off.

In the organ systems we explored in medical school we saw many examples of biological negative feedback. These are protective mechanisms that keep us on an even keel (in medical language this is called *homeostasis*). It means that harmful changes are sensed and automatically counteracted – the reason for negative feedback is to maintain order and health. Let's demonstrate a couple of examples in humans. For us to function efficiently we need to be at the correct temperature and have the correct proportion of water in our bodies. Here's how negative feedback works to automatically regulate this.

Sweltering (Drip) . . . or F-F-F-Freezing

It is essential that we keep our own body temperature at around 37 °C. All the chemical reactions in our bodies rely on thermal motion (the continuous movement of our atoms) at a particular rate. This rate is set by our temperature. If our temperature goes up to 40 °C, then we get heatstroke; if it goes down to 35 °C we develop hypothermia.

Our own internal thermostat tries to control our body temperature to within quite a narrow range. We have all experienced getting too hot or too cold. What happens? Sensor says too hot: switch on coolant mode and start to sweat (when the sweat evaporates it cools the body by taking heat). Sensor says too cold: switch on heating mode and start to shiver (the muscular activity of shivering produces internal body heat).

Thirsty?

Another example of negative feedback is our hydration system. Once we understand how our body regulates its water content, it becomes easy to understand how it also regulates its energy content, and therefore how much fat is stored – the hydration and energy storage systems are similar. All doctors know how we regulate water in our bodies – this *is* taught in medical school – but I imagine that only a minority of them grasp energy regulation.

Let's look at the hydration system. This negative feedback system has one sensor connected to two switches. Water makes up 70 per cent of our bodies. Beneath our skin, we are basically immersed in a 37 °C salt-bath. We need to make sure the water in our bodies is not too concentrated or diluted. If we become over-hydrated it can lead to seizures (and eventual death), and if we become too dehydrated we become weak and dizzy (and also, in severe cases, die).

The Sensor – the Kidney

The sensor to detect dehydration or over-hydration in the blood is in the kidney. Once it senses a change it secretes a hormone (called renin) which leads to a message being sent to the two switches. The two switches control:

1. The amount of water we take in – by controlling our thirst
2. The amount of water we let out – by controlling how much urine we make.

We Only Need 700cc But are Thirsty for More

The kidneys need to purify the blood of waste (urea) by producing urine. They can do this by producing just 700cc per day.* If we excrete

---

* In critically ill patients, the minimal urine output should be 30ml per hour to prevent kidney failure and to ensure survival. This equates to 700ml per day. We

below this volume of urine, we become unwell and start to develop kidney failure, so the kidneys signal for us to drink about double the minimum amount of water needed for health. We therefore drink about 1.5 litres of water per day and produce the same amount of urine. We don't need to drink 1.5 litres – we could survive on about 700ml per day – but as an insurance mechanism our thirst switch is ratcheted up so that we have plenty of essential water going through our system.

Biological systems like to be on the safe side, so, in this case, they habituate us to drink much more water than needed. Biology likes a safety buffer – this is an important point to remember when we compare our water-regulating system to our energy-regulating system. If we go without anything to drink for a few hours, then the kidney senses this. It sends a signal to turn on the switch located in the brain that controls thirst – the water-in switch. The brain gets the thirst signal and all you can think about is getting water. The more dehydrated you are, the stronger the thirst signal. At the same time the kidney sends a signal to turn off the water-out switch. We then produce only the minimal amount of concentrated, dark urine – less water is excreted and more is retained. Dehydration fixed.

The sensor also works the other way around, so that if you have drunk too much water and the blood is over-hydrated, it will turn off the first signal to the brain and you will not want to drink any more. It also flicks on switch number 2 in the kidney, leading to lots of dilute urine being produced. Less water in and more water out – over-hydration corrected.

Counting Calorie Intake? We Never Count Water Intake!

This negative feedback system works constantly to regulate the amount of water that we have in our bodies. It works subconsciously. In a whole year, we drink over 550 litres of fluid. That's the equivalent of five full bathtubs of water passing through our bodies every year. But we never have to measure that water to make sure that we are drinking the correct amount. Doctors don't have to warn us that if we take in 6 litres of water more than we excrete we could die of over-hydration – they know it is powerfully regulated without

---

also lose water through our breath (400ml), through sweating (400ml) and in faeces (100ml), but this is offset by the amount of water we generate through our own metabolism (400ml) and the water contained within the food we eat (500ml).

having to think about it. We don't have a 'water in – water out = water stored' equation in our minds. This is because we know that our water balance is controlled by our biological negative feedback mechanism. And the mechanism is exquisitely accurate. Of the 550 litres consumed per year an identical amount is lost from our bodies, all without a conscious thought.

People do occasionally die from drinking too much water (6 litres in a short period of time), but they are consciously over-hydrating. Rare examples are: inexperienced runners in a marathon, who fear dehydration and therefore force themselves to drink too much, or young students playing drinking games. Either can be rapidly fatal.

Just like the hydration systems, energy metabolism (i.e. the amount of energy taken in, the amount used and the amount stored) is critical for the survival of any species. All species go through times of feast and famine; the ones that survive and thrive are the ones that can predict exactly how much energy may be needed and should be stored for the future.

Six Big Macs . . . with Six Sides of Fries . . . and Six Cokes

Let's go back to Metabology Rule 1. This is the rule that most people use to understand obesity: (Energy In) – (Energy Out) = Energy Stored. Scientists have calculated that to store 1kg (2lb) of fat you need to take in 7,000 extra kcal.[1] That's the equivalent of six Big Macs, six sides of fries and six Cokes – on top of the usual calories that are needed every day. So, fit in a Big Mac meal on top of your normal meals for a week (excluding Sunday) and you will gain 1kg or about 2lb.

The traditional explanation for the overwhelming rise in obesity in the last thirty years is that we have been consuming too many delicious Western-type foods, too many Big Mac meals. On top of this, we have more cars, dishwashers, video games etc. and therefore do not move around as much as we used to. Basically, the conventional wisdom is that we have created a society in which it is easy to become too greedy and too lazy and this has led to us getting fat. It's our fault. If we just take Metabology Rule 1 to explain obesity, then this conclusion must be correct.

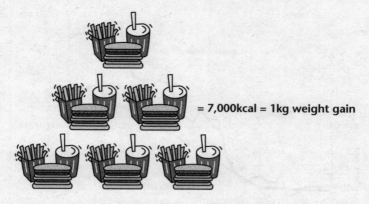

Figure 1.2 7,000kcal translate to 1kg of weight gain

Why aren't All Americans over 300kg?

If we look at the data, it seems that this conclusion is correct. The rise in the rates of obesity started in the early 1980s and this seemed to coincide with the rise in the consumption of calories by the population. In fact, if you look at the statistics from the US, the rise in the calories in the food supply exactly corresponded to the rise in obesity rates.*[2] In 1980, the average American man consumed 2,200kcal per day. By 2000 he was consuming 2,700kcal per day.[3] In 1990, he weighed 82kg (12 stone 12lb) and twelve years later the average American man weighed in at 88kg (13 stone 12lb). The data seems to back up the traditional theory of obesity – that it is a simple energy-in/energy-out equation. But there's more to the story.

So, at first glance it seems to be clear: calories cause obesity. But hang on, if we look at the figures more carefully, they don't add up. The average American man is eating 500kcal more per *day* during this period. What's that per year? 500 x 365 = 182,500kcal extra. How much weight should the average American man gain per year if we use Metabology Rule 1?

---

* Calories taken in from the food supply, when adjusted for food wastage, are the most accurate method of determining a population's calorie consumption. Several studies have used self-reporting of food intake to estimate consumption. This was recently confirmed by the UK's Office for National Statistics to be up to 70 per cent inaccurate.

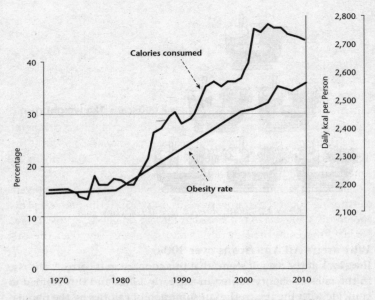

*Figure 1.3 Obesity rates started to take off in 1980, coinciding
with the rise in calories consumed*

Source: C. L. Ogden and M. D. Carroll (2008). Prevalence of Overweight, Obesity,
and Extreme Obesity Among Adults: United States, Trends 1960–1962 Through
2007–2008. *National Health and Nutrition Examination Survey (NHANES)*, June.
National Center for Health Statistics.

If we assume that the amount of physical activity wasn't increased,
and there is certainly no evidence of this, then applying our rule over
a year leads us to the following conclusion:

500kcal per day over one year:
extra energy in – extra energy out = energy stored
182,500kcal – 0kcal = 182,500kcal
1kg fat = 7,000 extra kcal
Expected weight gain over one year = 182,500 / 7,000
$$= 26kg (4 stone)$$

A predicted 26kg weight gain in one year. In twelve years, the weight
gain of the average American man would be 312kg (49 stone)! But
the actual figures say that in this period the average American man

gained 6kg (13lb) in total (or 0.5kg per year, not 26kg per year). What has happened to Metabology Rule 1?

This takes me back to my first few visits to the US, usually for conferences or to teach surgery. When you first visit, everything seems bigger, including the people. I observed the portion size and the type of foods that Americans eat. I went to their gas stations and supermarkets and saw how everything had been supersized, with tremendous amounts of sugar and fat added to their foods. My thought at the time was, 'Why aren't Americans even larger?' Looking at the figures now – 182,500 extra calories per year – I wonder again why all Americans do not weigh 300kg.

The actual weight gain for Americans who, as a population, were consuming 500kcal extra per day was only 0.5kg (1lb) per year. This equates to 3,500kcal extra, stored as fat, over the year, or just 11kcal per day, the equivalent of one potato crisp, per day, over the calorie limit. Not one packet of crisps, but one crisp. This means that although the average American is consuming so much more than necessary, they regulate their energy balance to within 0.4 per cent of perfection. A separate validation study that more accurately measured energy usage over a year and weight gain found the system to be even more accurate, with only 0.2 per cent of calories ingested being stored as fat.[4]

What happened to the 'missing energy' of 489kcal per day? To answer this, we need to go back to the rule that is often ignored when explaining obesity: negative feedback.

Hoarding Energy

Remember that the negative feedback rule is designed to protect the body against unhealthy changes – by activating processes that will oppose those changes. We know that there are many of these types of mechanisms at work in the body, helping to preserve a healthy state. The regulation of our temperature and hydration are just two of these systems. We know that energy regulation and storage is a critical part of survival in animals. You need to store energy for times of need, but you cannot hoard energy indefinitely, because if you do, as with any hoarding behaviour, things get messy and there is no room to move. So, we should not be surprised if the amount of energy stored within our bodies (just like the amount of water) is also controlled by a negative feedback mechanism. This would explain why, in the presence of so much over-consumption of food, American men's weight edged up by much less than predicted.

But how could a negative feedback system work to stop massive weight gain? We know the energy has gone into the body, but the body has not stored it. Therefore it must have been used up somehow. But where? Let's recap energy expenditure:

Energy expenditure = Active energy expenditure (gym)
+ Passive energy expenditure (walking/moving)
+ Basal metabolic rate (breathing/ heartbeat/temperature control)

How is the extra energy used up? Do people sense they need to exercise when they over-eat? Most people give it a few seconds' thought but don't act on it, so we can discount *active energy expenditure* as the most likely scenario. Some scientists say that people might fidget more when they over-eat and this uses up the extra energy in the form of *passive energy expenditure*.[5] But to use up nearly 500kcal per day just twitching your legs is a lot of twitching considering that walking for a mile uses less than 100kcal. I don't think we fidget our way through that much energy. What about the *basal metabolic rate*? Does the body ratchet this up in order to stop us storing too much energy?

The Vermont Prison Feast

To start to answer this question we must go back fifty years to an extraordinary experiment.[6] A team of American scientists, led by Ethan Sims, set up their lab in Vermont State Prison in Burlington, Vermont. They were studying obesity and wanted to observe and analyse what happened when a group of men deliberately over-ate to increase their body weight by 25 per cent over a three-month period. Over-eating takes time and needs to be supervised. The scientists had started the study using students, but aborted it as the students did not have enough time, between their studies, for supervised over-eating. Prisoners were much more suitable for the study. They had nothing else to do and their activity could be monitored (they were to be barred from physical exercise). The scientists negotiated the promise of early releases for the prisoners who were able to gain enough weight to meet their target.

The scientists employed a dedicated chef for the prisoners and upgraded their plates from tin to china. Breakfast was a full American: eggs, hash browns, bacon and toast. Lunch was unlimited sandwiches. Evening meal was steak or chicken with potatoes and veg. Before bed they sneaked in another full American breakfast. The men started out by increasing their calorie consumption from 2,200kcal to 4,000kcal per day. The scientists observed a steady rise in the weight of the prisoners, but then a strange thing happened that puzzled the scientists. Despite eating 4,000kcal/day, the prisoners stopped gaining weight. They could not put on any more and were still a long way short of their target of a 25 per cent weight increase.

2,200 to 4,000 . . . to 10,000kcal

So the scientists ratcheted up the calories. Most men had to eat 8,000–10,000 calories per day to keep putting on weight, four times what the scientists had calculated would be required. Astonishingly, a few of the prisoners seemed resistant to further weight gain, even at 10,000kcal. Why could they not put on any more weight? The answer came when the scientists measured the metabolic rates of the over-fed, and now overweight, prisoners. In all cases their metabolism had increased considerably. The men seemed to be adapting to the over-eating environment by burning off more energy in order to protect themselves from runaway weight gain. Does this sound familiar? It may explain why our average American male put on 6kg rather than the 200kg-plus that we had calculated from the increased consumption of processed foods in the 1980s and 1990s.

In 1995, a research group from Rockefeller University Hospital in New York investigated the effects of a 10 per cent weight gain on two groups of patients.[7] One group started with a normal weight and the other group were obese. Interestingly, the obese group had a higher than predicted resting metabolic rate than the non-obese group at the onset of the study, before any weight gain. A high-calorie drink made up of protein, fat and carbs was used to drive weight up. This helped the scientists calculate more precisely how much energy was being taken in. What happened to their energy expenditure when the two groups made the 10 per cent weight-gain target? As with the Vermont Prison study, the basal metabolic rates of all the subjects in the Rockefeller study increased – in the non-obese group by over 600kcal/day and in the obese group by even more, over 800kcal/day.

In a later study in 2006, researchers at the Mayo Clinic in Rochester, Minnesota, analysed twenty-one previous over-feeding experiments, including their own.[8] They confirmed that on average the basal metabolic rate did indeed rise by an average of 10 per cent in response to the over-feeding. The more energy that was taken in by over-feeding, the more the body tried to burn up those extra calories to stave off weight gain.

More Firewood – More Fire

These over-feeding studies suggest that, yes, there is indeed a negative feedback mechanism controlling our weight and stopping us gaining too much weight too fast. Imagine that you have a log fire at home. Every winter's day you have one log of wood delivered and every evening you relax by the fire and burn that piece of wood. Imagine, now, that you receive three logs of wood each day. What would you naturally do? You may not have much space to store the wood, so you would probably burn the excess, keep warm, have more energy and avoid the chill.

The scientific evidence that we compensate after over-eating by burning off more calories is compelling – and it fits in with our epidemiological evidence: we don't gain 26kg (4 stone) per year, just 0.5kg (1lb). But, if you ask most dieticians or doctors if they are aware of this mechanism – metabolic adaptation to over-eating – they will say no. This is not covered in their training. Why not? You would expect that something so fundamentally important would be understood by the medical profession, and should be accepted public knowledge.

Some scientists still argue that the increased energy expenditure that we see when we put on weight is because the body has become physically larger. A larger body burns more energy. However, when we analyse the figures this theory doesn't add up. Most people who gain weight, especially in over-feeding experiments, but also in everyday life, put on the excess weight as fat and not muscle. Fat expends a minimal amount of energy; compared to muscle it is a very efficient organ. In the Vermont study the prisoners had to consume 50 per cent more calories than expected just to maintain their increased body weight. Because their metabolisms were so 'hot', they all lost their extra weight as early as twelve weeks after the experiment ended and they were able to resume normal eating. None of them needed any type of diet to get back to their normal pre-study weight.

A study from Arizona, looking at fourteen men who over-ate 100 per cent more calories than normal, found that within the first forty-eight hours of over-eating (i.e. before any significant weight gain had occurred) their BMR had increased by an average of 350kcal per day.[9] The conclusion? Over-eating leads to the burning of energy through an increasing metabolic rate. When we compare how most of our organ systems are kept in check by negative feedback, then it should come as no surprise that there is some sort of negative feedback to protect us from storing too many calories.

Are our bodies trying to protect us from ourselves by burning more energy when we take in too much food, in a similar way that our kidneys rid us of excess fluid when we drink too much? This would explain why some people seem to be resistant to excessive weight gain despite eating far too many calories.

But here is an important issue raised by Metabology Rule 2. If the negative feedback mechanism is working to stop some people gaining as much weight as predicted, then it should also be working to stop people losing weight when they go on a diet. Could this explain why diets often fail?

'I Can Lose Weight, but I Can't Keep It Off!'

I hear this statement in every clinic that I have worked in. At least one patient will have said it in every clinic, every week, every month over the fifteen years that I have been seeing patients struggling with their weight control. Sometimes I tell the medical students who sit in on the clinic that my next patient will tell us this, and almost all patients prove me correct. Here is a typical example:

I have been dieting since my teenage years. I have tried all the diets out there. Weight Watchers, Slimming World, LighterLife, the red and green diet, the cabbage soup diet. I have tried them all.* I can lose weight but I can't keep it off. I can lose 5 or 10kg (1–1½ stone) on a diet but then after two or three or four weeks the weight loss stops. I'm still on the diet, I'm still counting my calories and starving hungry and tired and irritable, but after a while the diet doesn't seem to work any more. When I go and see my doctor and tell him that the diet is no longer working, he tells me that it is impossible and that I must be sneaking extra food in. He basically doesn't

---

* In chapter 12 of this book, we will look at the most common diets, how they work, and why they fail.

believe me. So, I stop the diet and the weight piles back on . . . fast. Usually I regain all of the weight that I had lost and then gain even more!

This is the classic story that I have heard many, many times in my clinic. But it does not correspond with the simple 'calorie in and calorie out' rule. It's difficult to understand why someone can restrict calories, sometimes to 1,200 calories a day, and after a while they stop losing weight.

Let's see what happens if we apply the same type of system that maintains our bodies' hydration – our negative feedback system – to weight control and our energy storage, i.e. our fat. Let's apply Metabology Rule 2. If the system mirrors our hydration system – and we know all biological systems work in a similar way, so this is likely – then there will be one sensor and two switches.

The sensor will detect the amount of energy stored in the body as fat. Once it senses a change in the amount of fat stored, whether it goes up or down, it secretes a hormone which leads to a message being sent to the two switches. The two switches control:

1. The amount of energy we take in – by controlling our appetite
2. The amount of energy we use up – by controlling our basal metabolic rate.

If the energy storage system in our bodies is really like our hydration system, then it will direct more energy intake than we really need. Remember, we can just about survive on 700cc of water/liquid per day but our hydration system wants us to drink 1,500cc.

The insurance mechanism built into our bodies tells us to drink double the amount of water than the minimum required for survival. Biological systems like to be on the safe side, so they habituate us to drink much more water than needed. In the same way, maybe our energy-regulating system directs us to consume more calories than we need and then burns off the excess. This would also mean that, when we calorie-restrict, it is all too easy for the body to cope with this. It would be similar, in the hydration system, to consuming 1 litre of fluid per day and not the recommended 1.5 or 2 litres. You would be able to survive indefinitely on 1 litre of water per day, but your biological feedback system would be screaming for more fluid by giving you a raging thirst and reducing the amount of urine excreted to a minimum. You would survive, but feel pretty terrible.

Does a similar thing happen with our energy regulation when we go on a diet?

Let's look at the evidence that our bodies adapt to calorie-restricted diets in a similar way they do to fluid restriction.

The Minnesota Starvation Experiment

In 1944 researchers at Minnesota University, led by Ancel Keys, an up-and-coming young scientist in nutrition, set up a study to look at what happens to people's metabolism during starvation.[10] The Second World War was coming to an end and the US recognized that millions of Europeans could be facing famine. They wanted to know the best diet to keep them alive. The Minnesota Starvation Experiment, as the study became known, recruited thirty-six male volunteers, who were conscientious objectors but wanted to help the war and subsequent peace efforts. They signed up to be confined to an allocated living area within the university football stadium, from which they were observed for one year.

The scientists first monitored them for twelve weeks while they ate a normal diet (the 3,200kcal/day stated in the study seems excessive, but the subjects were doing manual work). They were then put on a calorie-restricted diet of about 1,500kcal per day for twenty-four weeks while still doing manual work, and their weight, mood and metabolic rates were measured. After the diet period they were then observed for a further twenty-four weeks on an unrestricted diet.

As planned, during the twenty-four-week diet the subjects lost about 25 per cent of their weight. However, the scientists noted that their metabolism had plummeted by more than would be explained by the reduction in the size of their bodies. Their BMRs decreased by on average a massive 50 per cent of the starting value. Half of this value, 25 per cent, was unexplained by the change in the size of the subjects (smaller people have lower BMRs than larger people). It was as if their bodies were trying to adapt to the starvation environment they found themselves in – by shutting down energy expenditure to the absolute bare minimum. Their heartbeat and breathing were slow and their body temperature was low.

When the group resumed eating normally, their weights increased much more rapidly than would have been expected for their size. The scientists attributed the fast weight gain to the sluggish metabolism that the enforced diet had produced. In all the subjects the weight regain exceeded their initial weight at the start of the study. All the subjects ended up heavier than when they started the study. The distribution of their weight had also changed: they had lost some muscle mass and not regained it. All the weight regain was in the form of fat deposits (these outcomes may seem familiar to readers who have tried extreme diets).

Of interest to anyone who has tried dieting was the report of the psychological changes that the enforced diet brought to the subjects. They suffered with depression and anxiety, and had difficulty concentrating. They became hypochondriacs, worried about their health and wellbeing. They fantasized throughout the day and night about high-calorie foods. They lost their libido. One of the subjects became so depressed that he was reported to have chopped off three of his fingers with an axe. Many recurrent dieters will sympathize with the psychological craziness brought on by dieting. The Minnesota Starvation Experiment was the first study to prove that when you restrict a person's calories, they will respond, or adapt, by reducing their BMR. Less energy in leads to less energy out.

More recent studies have confirmed these phenomena.[11] Professor Rudy Leibel and his team at the Institute of Human Nutrition, Columbia University (and formerly Rockefeller University, New York), have been researching the changes in metabolic rate with dieting and over-eating since the mid-1980s. One of his lab's seminal studies recruited students to live within the hospital for periods ranging from three months to two years (I hope they got good grades after this). He studied, in great detail and using new techniques to measure metabolism accurately, what happened to the metabolic rate if a person over-ate and gained 10 per cent of their weight; or dieted and lost 10 per cent of their weight; or dieted for longer until they lost 20 per cent of their weight. Each test of metabolism cost the lab $500, so the experiment was expensive to run and therefore has not been repeated by other labs. Leibel found that when a person over-ate and gained 10 per cent of their weight their metabolism increased by 500kcal/day, just as in the Vermont Prison over-eating experiments. When his students lost this weight, and continued to lose weight until they were 10 per cent lighter than their initial body weight, he found that their BMR decreased by 15 per cent (or about 250kcal/day), more than could be explained by weight loss alone. This suggested that the body's reaction to calorie restriction is to decrease the amount of energy expended, just as in the Minnesota Starvation

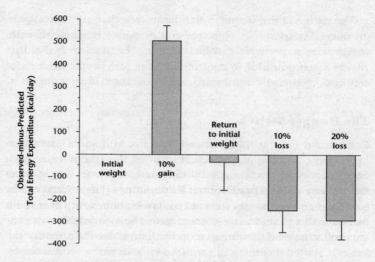

Figure 1.4 Unpredicted changes in metabolism after weight gain and weight loss

Source: R. Leibel et al. (1995). Changes in energy expenditure resulting from altered body weight. *N Eng J Med*, 332(10), March, 621–8.

Experiment and as we would predict if energy regulation has a natural negative feedback mechanism to stop runaway weight gain or weight loss. When Leibel measured metabolism after a 20 per cent weight loss, he saw only a modest further decrease to 300kcal/day. It was as if the body's protective mechanism, the negative feedback switch, was activated at 10 per cent weight loss.

All the studies on over-eating and under-eating had to be conducted in an enclosed environment, and were difficult to complete because the volunteers had to give up their normal lives for long periods of time. It was therefore difficult to recruit enough subjects and for this reason these types of studies are few and far between – and rarely quoted.

There are many studies observing the short-term effects of dieting on metabolism, but they are not relevant in our quest to explain the experience of real dieters. The long-term studies go part of the way to validating, and explaining scientifically, what dieters are experiencing. But only half the way, because so far we have only talked about one of the two switches in the negative feedback mechanism of weight regulation: the metabolism switch.

The nature of the second switch has meant that, until now, only prisoners, conscientious objectors and desperate research students would agree to participate in such studies. The problem is that this switch is too powerful to control and therefore people need to be confined, or virtually imprisoned, to prevent them from acting on it.

The Hunger Switch

One of the most striking elements of the Minnesota Starvation Experiment was the psychological changes the volunteers experienced as they lost weight and hunger overcame them. The subjects lost interest in their passions and surroundings. They would obsess over food and constantly stare at cookbooks, fantasizing over them like some sort of alternative pornography. The men would grow anxious and irritated if their small ration of food arrived late. One of the subjects started dreaming of cannibalism and, when confronted by the lead scientist about sneakily buying food during an authorized excursion, threatened to kill him. He was immediately released from the study and transferred to the psychiatric ward, but made a rapid recovery from his breakdown when he was fed normally for a couple of days.

Hunger is probably an even more powerful switch than metabolism when our bodies try and protect us from weight loss. We now know that the hunger switch is in the part of the brain controlling our body weight. It is a small pea-sized area at the base of the brain, just behind our eyes, called the *hypothalamus*. But its small size should not be misleading – it contains the switches of powerful basic needs, including the ability to produce a desperate thirst and a voracious appetite. The power of these two switches should not be underestimated. They will drive a human to extremes of dangerous behaviour to secure their goals of water or energy for the body. Most people who live in the developed world only experience hunger when they go on a voluntary diet. I am amazed by the self-control of some of my patients who can at times starve themselves for weeks on end. The strength of the hunger signal can be the same as that of a parching thirst for someone who has lost a considerable amount of weight. Its psychological effects can control your life. If you are in an environment where images and adverts and smells of delicious high-calorie foods surround you, there is only going to be one winner in the diet game – your hunger.

CASE STUDY – HUNGER WILL ALWAYS PREVAIL

Several years ago, a unit at a teaching hospital treated two teenage patients with exactly the same condition. They had tumours of the pituitary gland, the pea-sized gland that controls our hunger and thirst. Both of the patients, who were in their late teens, had undergone brain surgery to remove the tumour before it grew so big that it would put pressure on the optic nerve and cause blindness. The teenagers were of normal weight before their brain surgery, but following it the pituitary gland didn't work properly. It was unable to switch the hunger signal off. No matter how much they ate, they still felt ravenously hungry. They gained weight rapidly until eventually they found themselves at the doors of a metabolic surgery unit. Weighing 180kg and 200kg (28 and just over 31 stone), they were both deemed sensible enough, motivated enough and fit enough to have an operation (called the sleeve gastrectomy) to make their stomach smaller. The procedure was to make the stomach into the shape of a tube or sleeve and reduce its size drastically. They both had successful operations and lost considerable amounts of weight within the first year of surgery. However, their pituitary switch remained untreated – their hunger drive had not been changed by the surgery, only the size of their stomachs. The boy who had weighed 200kg had confessed before surgery to sometimes eating a box of crisps (a forty-packet variety box). Unfortunately, after one year, when less supervised by his parents, his appetite had driven him back to his box-a-day habit. Both of our patients who had lost weight regained all of it within two years, despite having a very small stomach capacity. The hunger drive really can overcome anything.

The major studies that I have described, looking at weight gain and weight loss and their effect on metabolism, are rare. It's difficult to recruit subjects for such arduous and long-lasting studies when ravenous hunger, or nauseous satiety, must be faced, and that's why these studies of human volunteers are so few and far between.

However, there are many animal studies that confirm the presence of metabolic adaptation to over- and under-eating. Negative feedback operates in many species to protect against extremes of weight gain or weight loss.

In 1990, a breakthrough in our understanding of metabolic adaptation came when scientists discovered a hormone that was produced by the fat cells and seemed to work on the hypothalamus to switch hunger and metabolism on and off. Finally, we had the last piece in the jigsaw that would prove the existence of negative metabolic feedback. The hormone is called *leptin*.

The Fat Controller

Leptin is released by fat cells – not in response to any signal, it is just released. That means that the more fat you have, the more leptin there is in your blood. Leptin is *the* signal to tell the hypothalamus how much fat we are carrying; it's like the petrol gauge in your car telling you how much further you can go, how much energy there is in the tank.

With the discovery that fat produces the messenger hormone leptin, we now have a negative feedback system for energy expenditure that, as predicted, looks remarkably like the negative feedback mechanism for hydration. The signal comes from fat, via leptin, and the two switches for hunger and metabolism that control energy in and energy out are in the hypothalamus.

Leptin works like this. After a period of over-eating, the volume of fat increases. Leptin is produced within the fat cells and goes directly into the bloodstream. The hypothalamus (weight-control centre in the brain) reads the leptin message and realizes that there are adequate energy stores and that no more is needed. It then acts by decreasing appetite and increasing both the feeling of satiety (therefore decreasing the amount of energy taken in) and the body's metabolic rate (increasing the amount of energy burned). These factors act to keep weight within a predetermined range (see the description of the weight set-point below).

Leptin also acts powerfully to stop weight loss. When weight is lost after a diet (or famine/sickness), the amount of fat available is reduced. This means that leptin levels in the bloodstream decrease. The hypothalamus senses this and acts to stop further loss of energy by increasing appetite, decreasing satiety (increasing energy in) and reducing resting metabolism (decreasing energy out). These actions slow down or stop further weight loss. When food becomes freely available once more, weight will be gained. This system explains how many people seem to be able to seamlessly control their weight for years and decades, without dieting or counting calories.

But there is one problem with this system. It does not explain why some people become obese. If the system worked perfectly, then obesity would not be such a problem. We have accepted that the system works *almost* to perfection: there is a 0.2 per cent error. Yes, 0.2 per cent of the excess calories that we consume, on average over a whole population, are stored and not used up. But if the negative feedback system is so powerful that it can alter metabolism up or down by 25 per cent and seriously affect the amount of food we take in by altering our hunger drives, then why is it not 100 per cent effective? Why does it differ in this respect from the hydration system which is always accurate to 100 per cent and gives perfect control of water balance within our body for a lifetime? There must be a biological explanation for this.

Fat Storage is Calculated
Let's think laterally about this, because it doesn't make sense that a biological system almost works. Let's assume that it does work to 100 per cent efficiency, but that the brain has made the decision to store more fat. The brain has sensed, from the environment, that it would be in its best interest to carry more fat. To clarify, the negative feed-back system for energy (fat) storage is working to perfection, but the brain has calculated, based on the incoming data from the environment, that it needs to increase the stored energy reserves. We would expect that it has made this decision using information from the past as well as the present to predict future energy needs – it may even be using genetic data passed from previous generations.

Preparation for a Famine
Why would our brain calculate that it would be safer if we carried more energy? Why does it want a bigger fuel tank? The most obvious explanation is that it senses that food may become scarce in the future; it senses a famine or a long harsh winter is on the way. Maybe it has received signals in the past of major food shortages (historically a famine, but in the present day more likely a low-calorie diet). It logs these experiences and calculates that, to be on the safe side, we may need a little more fat just in case the next food shortage is worse. Or maybe it has sensed that the quality of the food in the environment

is similar to that found in the autumn and it's time to tell the body to store more calories for the winter – just like a brown bear which will automatically develop a voracious appetite and increase its body weight by 30 per cent in a few short weeks before hibernating, in response to cues from its environment.[12]

Our energy storage is too important to be left to free will. Although it appears that the amount we eat is under conscious control, in actual fact, it is our subconscious brain that controls our underlying hunger and eating behaviour. If the brain wants more energy on board, it will signal more hunger and less metabolic waste, and our weight will go up.

Suggesting that energy storage is under conscious control because we can deliberately stop eating for a period of time is like suggesting breathing is under conscious control because we have an ability to hold our breath. We don't have to remember to breathe – our subconscious brain does that for us. If we change environments and live on a mountain with thin air, we don't have to tell our brain that we need to breathe more quickly or deeply – the subconscious brain will sense the environmental change and breathe more deeply for us. In the same way, I think that for some of us certain environmental signals (such as an impending famine or a long winter, as we will see later) lead the brain to want to store more fat.

The Weight Set-Point

The level of energy (fat) storage that our brain calculates is necessary for our survival is called our *weight set-point*.[13] This is like the thermostat that controls the temperature in our house. It will reach, and then maintain, the level that it is set at, by using negative feedback systems.

The weight set-point is the king of Metabology Rules 1 and 2 – it drives them. If your weight is less than your weight set-point (maybe you have been sick or on a diet), then Metabology Rule 2 (negative feedback) will kick in and you will be directed to eat more and your metabolism will shut down. Metabology Rule 1 will then act to shift the weight back up (more energy in + less energy out = energy storage). Likewise, if our weight is above the set-point (maybe we have been over-eating on holiday), then Rule 2 will direct us to eat less and at the same time our metabolism will increase – Metabology Rule 1: less energy in + more energy out = weight loss – until the weight set-point is achieved.

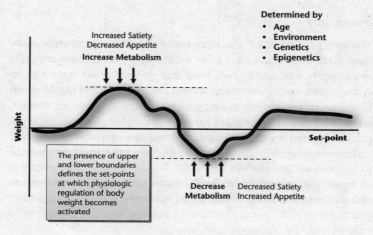

Determined by
- **Age**
- **Environment**
- **Genetics**
- **Epigenetics**

Increased Satiety
Decreased Appetite
Increase Metabolism

Weight

Set-point

The presence of upper and lower boundaries defines the set-points at which physiologic regulation of body weight becomes activated

Decrease Metabolism Decreased Satiety
Increased Appetite

Figure 1.5 The weight set-point

Unfortunately, the weight set-point is not always set at a healthy weight. If you suffer with obesity then your set-point is probably the cause of this. If you consciously try and lose weight by traditional means (i.e. eating less and moving more with no change to the quality of the food you eat), your powerful negative feedback mechanism will force your weight back up. It becomes a struggle of wills between your conscious desire to be a particular weight and your brain's subconscious power to regain its desired weight set-point. Invariably – but unfortunately for all dieters – biology always wins. It may take a week or a month or a year or even several years, but your subconscious brain will eventually haul your weight back to where it wants it to be.

The weight set-point theory, plus negative feedback to control your weight at that set-point, fits in with both sensible biological models and with what patients tell us of their experiences with obesity. They feel trapped, not in control. They can lose weight but will always regain it as the subconscious brain wins the battle of wills. If, by losing weight on a diet, you send signals of probable future famine to the brain, then not only will you regain the weight you lost, but invariably your set-point will edge upwards and you will end up heavier than before you started dieting.

Summary

The secret of successful and sustainable weight loss is to understand how our bodies regulate our body weight set-point. It is not as simple as energy in and energy out. Now that we know that the set-point is the master controller of our weight, we need to find out how our brains calculate where it is set. Various factors in our environment, our history and our family background determine our individual set-point – whether it will be set as slim, obese or somewhere in between.

Later chapters will discuss how to identify the signals (in the type of foods we eat and the way we live) that control our weight set-point. Once we understand these signals, our weight set-point, and therefore our weight, will finally be under our control.

TWO

The Sacred Cow

*How Genetics, Epigenetics and Our Food
Control Our Weight Set-Point*

I was sitting at a dusty roadside tea shop in rural India, daydreaming and watching the sun go down, when I noticed that the traffic had come to a complete standstill. The chaotic weaving of lorries, cars, bikes and tuk-tuks had ceased. Normally when this happens there follows a cacophony of horns as the source of the roadblock is abused, even if the jam happens to be caused by an accident. But this time there was unusual silence and I wondered what was going on. Five minutes passed and the traffic started up again, but at a slower pace than before. Then, calmly walking in the opposite direction to the traffic, in the middle of the road, I saw the source of the traffic jam, a cow, sacred and revered by Hindus.*

The animal looked fairly bedraggled in the heat and noise, but what made me curious was its size. I knew that it would have been very well fed by its owner, who had placed colourful garlands around its neck, but despite this it seemed slim and wiry compared to the cattle that I was used to seeing in the drizzly green fields of England. The cows that I remembered were double the size and the discrepancy puzzled me. Why would well-fed Indian cows remain slim, compared to the same animals in my country?

---

* Anyone who has travelled to India will have noticed that cows are allowed to roam the streets and roads freely. Early Hindu Vedic texts encouraged the peaceful coexistence of cows and man. Cows are venerated for their peaceful temperament and maternal qualities such as milk-giving – a very important food source for the population. Not only is their milk used, but the ghee (the clarified butter from the milk) is used for cooking and burned during blessings. Everything that the cow produces is seen as useful (its dung is used as fuel to burn in the winter or as fertilizer in the summer), and it is not unusual to find people in rural areas drinking and bathing in cow's urine (a useful source of sterile fluid).

The answer to this question is quite simple, but can lead to a better understanding of the factors in play with regard to humans: how our environment and our genetics determine our weight set-point. And when we truly understand this concept, we can beat obesity.

Imagine that you are a cattle farmer and you want to maximize your profits. How can you make your farm and your herd better than your neighbour's? How can you make your cattle grow bigger so that you can make greater profits when they are sold? The first and obvious answer would be to ensure that there is a plentiful food supply in both summer (grass) and winter (hay). This would surely optimize their size. All the cattle would be well nourished. But hang on, aren't the slim sacred cows in India well fed all year round? So maybe the obvious answer is not the correct one.

There are two common strategies used by farmers to make their cows grow larger than non-farm, or wild, cows. When these practices are applied to human populations, you've guessed it, obesity results. Here is why.

Drive-Through for Cows

The first strategy is not to feed the cows what they would normally eat and what generations of cows have eaten for millennia, i.e. grass. If you change the food that the cows are given from grass to a mixture of grains and vegetable oils, then you have a good chance of fattening them up so that they are worth much more when you come to sell them. We know that this happens quite regularly on commercial farms. Cows that consume a diet including grains, such as maize and soy, mixed with oils such as palm oil, will gain weight considerably faster than those grazing on grass.

To optimize weight gain even further and faster, farms introduced feeding lots so that the animals are confined to a pen and cannot really do anything but eat the corn or oil feed that is right in front of their noses for hours on end, for months at a time. The present-day human equivalent would be a kind of fast-food drive-through (with use of all the car's entertainment systems banned – but free food), where the car does not actually drive through but stays by the serving window. The recipients would be so bored that they would continue to eat the tasty high-carb, oily foods passing in front of them throughout the day, every day. Imagine being stuck in this situation for the decade that you were growing and you can imagine what would happen to your waistline.

It's not just cows that will grow bigger in response to a change in their diet. We know from studies of rodents that if you want them to grow faster, and fatter, you cannot do this by just feeding them more of their natural food (called chow). It is not the amount of food that will alter their weight set-point, it is the quality of the food. If a rodent is given high-calorie, high-fat foods (called 'canteen food' by the scientists), its weight set-point will be raised.[1]

So feeding cows a high-calorie grain-and-oil mix, and detaining them in a pen, makes them grow faster. This is not really rocket science, but the point is that this type of change in diet, towards more grain- and oil-based foods, when mimicked in the human population, causes a similar change in size – people will get bigger and fatter. It's the same with all mammals. On the whole, we are no different in our metabolic biology to those farm cows and lab mice and rats.

Once the food supply to a human population is changed to 'canteen'-type food, then many of that population will become obese. One of the interesting factors in this dietary change that I have noticed over the years is the real difficulty for most of the population to be able to purchase 'normal' fresh food (i.e. food that is not processed). When you leave the office in search of a healthy lunch, it can be very tricky to find foods that haven't been manipulated or changed into high-calorie food replacements. The Western high street is like a food desert – natural foods only exist in rare, difficult to find, oases. The mirage of real food is around, but it doesn't really exist.

Survival of the Fattest

Now we are ready for the second strategy that farmers can use to make their herd bigger and more profitable than their neighbour's. In every herd of cattle there are individual differences between animals. They are obviously not all identical. These individual differences (called *heterogeneity* in medical jargon) are very important for the survival of species. If some of the species are taller or shorter, or bigger or smaller, or faster or slower, the individuals at the extremes of the spectrum may be more likely to survive unexpected changes in the environment. For example, if there was a famine, the cows that were carrying more energy reserves (fat) before the start of the famine would be more likely to survive. Because more cows with a tendency to carry extra weight had survived the famine, the next generation of cows would be more likely to become fatter compared to the previous generation. In other words, this is an example of Charles Darwin's theory of natural selection or survival of the fittest (or, in this case, fattest).

Farmers can use the differences in characteristics between cows within a herd to make all the cows bigger and fatter by an artificially induced natural selection (or, more succinctly, *unnatural* selection, because it is not the natural environment that is selecting the next generation but the farmer). For instance, they will select those cows that put down fat deposits within their muscles to give the eventual meat the tasty marbling effect that you may know from a rib-eye steak. Obviously, this type of fatty meat is more valuable to the farmer. When full grown, the cows with these characteristics will be selected by the farmer for breeding the next generation; the ones that do not grow so big or so fat will not be selected and their 'slim' genes will be lost to the next generation of cows. If they continue with this unnatural selection generation after generation, then within ten generations the farmers who have used this method will have a herd of cows that grow much faster, are bigger and have more fatty muscle marbling, compared to the farmers who just concentrated on looking after, and feeding, their herd. This method of manipulating the gene pool of the herd to make them more likely to exhibit characteristics that are valuable to the farmer is known as *selective breeding*. It is the reason that, of the 1.4 billion cows on Earth, there are now over 1,000 different breeds, exhibiting favourable individual characteristics for the farmer.

Who Can Grow the Biggest Cow?

What can dietary manipulation and selective breeding in cows tell us about the human obesity crisis? Let us imagine three pens of cattle next to each other. The cattle from each pen are from three separate farms and each farm employs different farming practices:

- In the first pen the cattle are fed only grass and hay
- The second pen has cattle fed on canteen foods (corn and palm oil)
- The third pen contains cattle fed on canteen foods, but these cattle have also been subject to over ten generations of selective breeding to encourage rapid growth of fat-containing muscle.

How will the different pens compare?

Pen 1: Grass-Fed

The grass-fed cows will look similar to the sacred cow that I saw blocking the traffic in India – not much extra fat. As there was no selective breeding, there would be more differences between the cows, with some being bigger and some smaller, but most would be normal sized.

Pen 2: Corn-Fed

The corn-oil-fed cows would, on average, be significantly larger than the neighbouring grass-fed cows. Their weight set-point would be raised by their change in diet. However, as with Pen 1 cows, there will have been no selective breeding, so there would still be a significant difference in characteristics *within* this herd. Some of the cows at the lower end of their herd's size spectrum could easily be unrecognizably different if placed in Pen 1, despite eating a completely different diet throughout their lives.

Pen 3: Selective Breeding + Corn-Fed

The selectively bred and corn-fed cows would look massive compared to Pen 1's cows and would be much larger on average than Pen 2's. However, the larger cows in Pen 2 would not look out of place in Pen 3 despite never having been selectively bred (these are the cows that would have been chosen for selective breeding if it had happened in their farm).

If the differences between the cows in the three different pens were transferable to human characteristics, what would that tell us about the obesity crisis and who is affected?

It would suggest that if a group of humans were in an environment where they only consumed natural types of foods, then they would not really suffer significantly with an obesity problem. We'll call this *Group 1 Humans*.

If the humans were exposed to canteen-type food (i.e. grain/oil-based food with a high-calorie density), then this group would on average be much bigger and more obese than those groups that were eating natural foods. We'll call this *Group 2 Humans*.

Finally, if the group of humans had been selected to favour survival (and reproduction) of the biggest and most obese, and had also been fed canteen-type food, then they would on average be the biggest of the groups: *Group 3 Humans.*

So, is the farmed-cow model a good one to unpick the causes of human obesity? Let's look at the evidence in humans.

Hadza Hunters

It is very difficult to find a population of humans that still eats the same food as its distant ancestors did millennia ago. We know that there has been a massive shift in the type of foods available to 'Western' populations since the industrial food revolution of the last hundred years (more on this in chapters 7 and 8). However, the food that humans were accustomed to eating started to change around 20,000 years ago with the advent of agriculture, so we need to go back even further to the time when our ancestors ate only what they hunted and collected. Learning about the lives of hunter-gatherer populations is essential in order to understand who we are now, and how we reacted to our changed environment. There are a very select few nomadic, hunter-gatherer tribes remaining in the world today, among them isolated rainforest tribes in the Amazon, the Pygmies of the Congo jungle, the Bushmen of the Namibian desert and the Hadza people of the savannahs of Tanzania.

As part of my research for this book I was lucky enough to spend time with a Hadza tribe to gain first-hand knowledge of these unique people who represent humanity in its oldest and most untarnished form. The tribe that I got to know consisted of several family groups. The Hadza are a pure hunter-gatherer people and are proud of their culture and heritage. Visits from Western researchers do not dilute their lifestyle and they do not like to accept gifts or money; they prefer any funds and resources to be used to protect their land and their way of life from encroachment by farmers. You will not be surprised to learn that the Hadza tribe do not suffer with an obesity problem. They consume meat, berries, fruits, tubers (like sweet potato), and their favourite food is natural honey straight from the hive. These are the foods that they have been consuming for 150,000 years and they see no reason to change their lifestyle. They probably wonder why people would grow food like farmers do, when one can take it for free (in their case, directly from the savannah).

When you analyse the weights and sizes of individuals in the hunter-gatherer tribes you find a pattern common to all animal

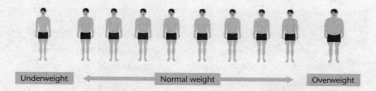

Underweight ← Normal weight → Overweight

*Figure 2.1 Frequency of underweight, normal-weight and overweight
people in hunter-gatherer tribes*

species which are consuming the natural foods that they were
evolved to eat (just like the cows in Pen 1). Some of the population
are underweight and a few are bigger than normal and overweight,
but the majority (80 per cent) are within the normal weight and size
range.[2] They have what statisticians would call a normal, or symmet-
rical, distribution to the size of their population. (See fact box below.)

This distribution of body types is the same for all types of animals living on natural foodstuffs, from chimps to lions to cows. Even
when there is an abundance of natural foods available to these animals, you don't see the population becoming obese. This suggests
that high-calorie availability will not affect weight, as long as those
calories come from natural foods.

FACT BOX: WHAT IS A HEALTHY WEIGHT?

Doctors and scientists commonly use a term called the Body Mass
Index (BMI) to calculate whether someone is underweight, overweight, obese or in the normal weight range. They cannot rely on
weight alone as body size depends on both weight and height.
For example, a 70kg (11 stone) woman who is 5ft 7in tall will
have a normal BMI and a 70kg woman who is 5ft 2in will have a
BMI in the overweight range.

BMI is calculated by dividing a person's weight in kilograms by
their height in metres squared. $BMI = kg/m^2$. The normal healthy
range of BMI is $18-25kg/m^2$. Someone is underweight if their
BMI is less than $18 kg/m^2$; overweight if their BMI is $25-30kg/m^2$.
Obesity is diagnosed if someone has a BMI of over $30kg/m^2$.
A BMI of over $40kg/m^2$ is termed morbid obesity ('morbid' means
'diseased' in medical language).

BMI is an important predictor of health. The higher the BMI (above the healthy range), the greater the risk of developing Type 2 diabetes, high blood pressure, high cholesterol (all of which cause heart disease) and cancer. People with a BMI of 38kg/m^2 or over will die on average seven years earlier than those in the healthy range.[3]

However, the BMI calculation can be an inaccurate predictor of disease risk if it fails to take into account someone's build. A body-builder (imagine Arnold Schwarzenegger in his prime) carries extremely heavy muscles and may have a very small amount of fat, but if you calculated his BMI he would invariably fall into the obese category (because muscle is so heavy). BMI is only accurate in someone who has a normal build (and this has not been defined). For instance, people of Asian descent have, on average, less muscle mass and therefore their BMI under-predicts their obesity risk. In their case a BMI of 28kg/m^2 is defined as obese.

So don't worry so much if you are a well-built person who is in the overweight BMI range – your weight is probably healthy . . . but *do* worry if you have a slim build and are in the overweight BMI range – you may already be at risk of obesity.

Farming Communities

What would happen to the Hadza tribe that I visited if the land that they relied on for their natural foods was taken over, and they were forced to become farmers? We know from fossil evidence at the time of the advent of farming that within a few generations the Hadza would probably become sicker and their height would decrease. The quality of their diet would suffer as they consumed more grains and had less food variety. But what about their weight? Well, if we look at the weights of early farming populations we can see that although most are still remarkably fit, many more of those populations are now overweight rather than underweight and a small number are now on the borderline of obesity.[4] In fact, if we look at the population weight curve we can see that some seem to be more affected by the change in the environment than others.

What if we alter the type of food available to the population from

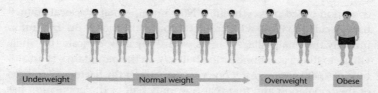

Figure 2.2 Frequency of underweight, normal-weight and overweight people in farming communities

hunter-gatherer food through farming produce and on to industrialized foods (the Western diet).* What happens to a population exposed to this type of food?

Among the UK population, approximately a quarter of adults are now obese; in the US this figure has reached 30–35 per cent; in the Gulf States, among the adult female population we now have obesity rates approaching 50 per cent.[5] On average, you could say that populations that are exposed to processed, or canteen-type, food are approximately one third normal weight, one third overweight and one third obese.

Is the Risk the Same for Everyone?

Does the change to processed food or a Western diet affect all people in the same way? Are we all at a higher risk of obesity or are some people more susceptible than others? Does the weight set-point of the whole population increase by the same amount or is there a difference in susceptibility between individuals?

If we are all at the same risk of developing obesity in response to a Western diet, then you would expect everyone in the population to be affected in a similar way. For example, take the population of a village living at the foot of a mountain in Switzerland. If you measure the level of haemoglobin (the blood test for anaemia) in the population, most people (about 90 per cent) will have a haemoglobin level in the normal range of 12–16g/dl. About 5 per cent of people will be anaemic and 5 per cent will have too much haemoglobin, a

---

* By industrialized foods I mean food that has been processed by food companies. The processing involves the removal of a lot of the goodness from the foods, to make them suitable for transport and storage, as well as to make them palatable, so that people buy them (in preference to fresh food) and the food company makes money. This type of food is what we mean by 'Western'-type foods.

condition called polycythaemia. Now imagine that the local council have decided to build a big tunnel through the mountain, but unfortunately the small village is in its path. They must relocate the whole village to living accommodation halfway up the mountain – at an altitude of 2,000 metres. After one year, they retest the haemoglobin level of the population and find that now only half of the population has a normal level and no one is anaemic. But half of the population has developed polycythaemia, the condition of too much haemoglobin (Hb). What has happened to the health of the population? Well, the much thinner air halfway up the mountain has caused the population to adapt by increasing the haemoglobin in the blood to compensate – haemoglobin carries oxygen from the lungs to our working organs, so if there is less air you need more Hb. But if you look at the distribution of the level of haemoglobin in the population, it looks the same as when they were living at sea level, it's just shifted up for everyone. The environmental change to thinner air has affected everyone equally.

In our population, if all were affected equally by the change in the food environment, then a similar scenario would occur to the Swiss villagers.

Figure 2.3 shows clearly that the majority of the population would be in the overweight category, with occasional members being of

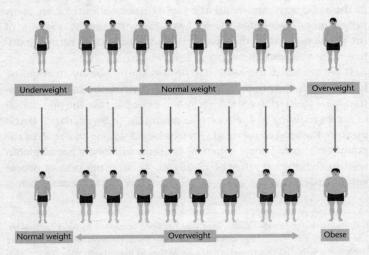

Figure 2.3 Change in population size if processed food affects everyone equally. Everyone's weight would shift upwards equally

normal weight and a similar small proportion obese. But this is not the case when we look at the current distribution of people's sizes.[6]

Some Immune, Some Hypersensitive

Some people, about one third, remain a normal weight and don't seem to have been affected too much by the environmental change. Another third have gone from normal weight to overweight and have been moderately affected by the change. But that leaves a third of the population who have gone from normal weight to become severely overweight (obese) – all because of the environmental change.

To make things simpler, we could separate anyone exposed to processed food as falling into one of three categories. These would be:

1. Obesity-resistant – still normal weight and able to maintain this weight easily.
2. Obesity-vulnerable – normal weight/overweight. Aware that if they consume too much processed food or don't go to the gym regularly they will gain weight.
3. Obesity highly sensitive – overweight/obese – struggle with weight even when trying to watch their calories and exercise.

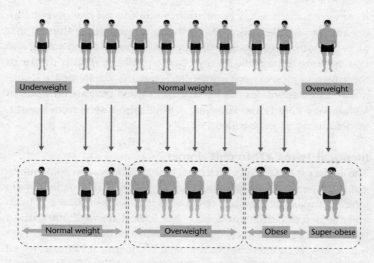

Figure 2.4 Actual change in weight of population after the switch to a Western diet. One third of the population is resistant to weight gain, one third is vulnerable to obesity and one third is highly sensitive to obesity

Free Will, Poor Education or Unlucky Genes?

The next question to ask when trying to understand obesity is which factors contribute to someone's sensitivity (or resistance) to developing obesity? Or, to put it another way, what leads someone to develop a higher weight set-point?

Is obesity a condition that is a person's free choice, as has been implied by most of the media for years, and supported by many of the scientists (we will consider the reason for this later), or does it come about because of home environment and parenting? Can we blame obesity in children on poor parenting? Or does it run in families, is it genetic? When I lecture medical students on this topic I ask them to rank in order the most to the least important factors in determining whether someone will become obese.

What would be your ranking of the most to least important factors determining someone's risk of becoming obese?

- Free will/character
- Home environment/parental influence
- Hereditary predisposition/genetic

If we were to ask the US population, we would get one overwhelming answer.[7] A 2012 poll of over a thousand Americans showed that 61 per cent of them thought that personal choices about eating and exercise were responsible for the obesity epidemic. This is similar to the answers I receive from my medical students – that obesity is controllable by free will (as you would expect when they are only taught Metabology Rule 1) and therefore anyone who suffers from it must, by definition, be weak-willed.

Identical Twins – Different Homes

The answer is in fact very different. Jane Wardle was an epidemiologist at University College London and published an eloquent study looking at pairs of identical twins who had been separated at birth and adopted into different homes.[8] She looked at over 2,000 pairs of twins and compared their BMIs (a measure of how obese a person is by considering their height and weight). Identical twins as we all know share identical DNA. They have the same eye colour, hair and complexion and are of almost identical heights.

What would happen to a pair of identical twins if one was brought up in a home with an unhealthy food and play environment (lots of processed convenience foods and not much outdoor play) and the other was brought up in a healthy environment? If the answer to our question was that home environment played an important role in triggering whether someone was going to be resistant to, or sensitive to, developing obesity, then we would expect these twins to have very different weights when they became adults. If the answer was that obesity was mostly inherited, then they would have similar weight when they became adults, despite not growing up knowing each other. If it was free will that determined size, we would then expect that their respective weights as adults would be fairly random and would not correlate with genetics or home environment.

The results may come as a surprise to many. The study found that there was around a 75 per cent consistency in the level of obesity (BMI) between identical twins when they became adults, despite their lifelong separation. She found that there was only a 10 per cent consistency in BMI as a result of the home environment.

This study firmly established that the main factor determining whether an individual would be normal weight, overweight or obese was not free will or upbringing but something that could not be changed by the individual – their genes.

Conclusive – Obesity Risk is Three Quarters Inherited

For the most part this study also exonerated parents, who had previously been criticized for poor parenting (i.e. they had been blamed for their offspring becoming obese). Only a small, 10 per cent, influence of home environment on obesity levels was found. Therefore if a child becomes obese it is 75 per cent because of their genes and 10 per cent influenced by their parental upbringing and home environment.

We should also highlight here the difference between the 'home environment' and the 'country environment'. If a country has adopted a Western culture, this trumps a healthy-living home environment. If Western culture pervades the home environment, the study indicated that, if you have obese-sensitive genes, even a healthy home environment may not protect you. The results of Jane Wardle's study of twins have been repeated and confirmed by many separate researchers around the world.[9] They certainly fit in with what many of my patients have been telling me for years – 'It's in my genes, doctor.' In this respect they are mostly right, but unfortunately this important research is still not widely known, or quoted.

Group 3 Humans

If we go back to our cow analogy, the largest and fattest cows were in Pen 3 and they were exposed to both canteen-type food and had a genetic predisposition to obesity (due to selective breeding). In humans who suffer with obesity we have now unpicked the two main factors involved in predetermining if they will struggle with their weight. These people would be in Group 3 Humans: a combination of the Westernized environment they live in (discussed in later chapters) and a genetic predisposition.

The genetic predisposition of some humans to become obese does not originate from artificial or unnatural selection, as in our cow analogy, but probably from natural selection. But are all human populations, races and tribes the same, or are some more susceptible to becoming obese – in the context of a Western environment – than others?

The Dubai Mall Food Court

I regularly travel to the Gulf to hold clinics for my patients and when I walk around their massive, air-conditioned shopping malls I am struck by how much more prone to obesity the local population of Emiratis seem to be compared to other ethnic groups living there, despite them having equal exposure to Westernized foods. The food courts are dripping with tasty hedonic Western foods from Burger King to Taco Bell to Subway. The mix of people within the food court's seating areas include Indian, Filipino, Caucasian, African and local Emirati. People from each individual ethnic group seem to be suffering with obesity, but the problem seems to be much more severe and more common in the local Emirati people. Is this backed up by research? A look at the current obesity league table confirms my observations: the Emiratis are near the top.

1. The Pacific Islands – the island of Nauru is top with 94 per cent overweight and 71 per cent of those obese (so only 6 per cent normal weight!).
2. The Gulf States (including Emiratis) – Qatar and Saudi Arabia are nearing 50 per cent obesity rates in female adults.
3. United States – 36 per cent obesity rate in Louisiana, with other states close behind.
4. Europe – 55 per cent overweight, 25 per cent obese.

The Pacific Islanders have stunningly high rates of obesity. It is almost as if they have been perfectly selected to have obese-sensitive genes. How could this be?

Brave Polynesians

As I write this, I turn to the globe on my desk, kept there so that I can daydream about the world and what's happening in it. Why are the Pacific Islanders suffering with such extreme obesity? The origins of the modern human are almost certainly to be found in East Africa, near to present-day Ethiopia. If I turn the globe around to the opposite side of the world to Ethiopia, we come . . . to the Pacific Islands.

Humans migrated from Africa and over many thousands of years inhabited all areas of the planet. Generations of tribes explored the Middle East, travelled through Asia and on to China. It is thought that Pacific Islanders originate from people living in what is now Taiwan and the Philippines. These people mastered the sea and finally discovered the pristine Pacific Islands. But here is a clue as to why Pacific Islanders are the biggest people in the world. The Islands are one of the last places on Earth to be inhabited by humans, probably around 1000 BC.[10] The distances involved in travelling by sea to these islands are phenomenal – thousands of miles. The Polynesian sailors would have followed the flight of migrating birds and used the stars to navigate. They probably gazed at the horizon for days and weeks on end looking for clues to nearby land – seabirds and turtles, twigs or coconut driftwood, or the distant build-up of cloud formations around an island. These journeys were the ancient equivalent of a moon landing, long and arduous and at the mercy of the unpredictable elements. Not surprisingly, many of the crew and passengers did not survive the journey, as described here in the collection edited by J. Terrell, *Von Den Steinen's Marquesan Myths*: 'The voyage was so long; food and water ran out. One hundred of the paddlers died; forty men remained. The voyagers finally reached Fitinui, then Aotona.'[11]

You can imagine the hardship and risk involved for those people to safely reach the Pacific Islands. Quite often only those people who were 'strong enough' to withstand the starvation of a long trip survived to live in these islands. There was therefore automatically a huge selection bias for anyone settled there. Those people who had enough fat reserves before the journey, or those with metabolisms that could shut down in the face of starvation, had a much better chance of surviving the long journey. The sailors and passengers who did not have this insurance perished and did not have the opportunity to pass their genes on to the next generation.

The massive selection bias of people with good fat reserves, or efficient metabolisms, surviving to colonize this distant part of the world is almost as extreme as the selective breeding of the cows in Pen 3 by farmers (see pages 35–6). In addition, once settled, these people were at the mercy of severe famines affecting their small isolated islands – migration to unaffected areas would have proved much more difficult than for those living in large continents. These famines again reselected for survival those people that had built up adequate fat reserves to keep them going.

Hidden Obesity Genes

The Pacific Islanders offer us a unique insight into how genetic selection can, in this case, favour survival of the fittest.[12] But for most of the history of the islands, from the early settlers to colonization by Europeans, the population was not overweight, because they had always consumed fresh natural foods. They were well nourished and able to withstand minor food shortages quite well, but there wasn't an obesity problem. It was not until the recent introduction of a Western-type diet into the islands that the time-bomb in the population's genetic make-up was unleashed. The Pacific Islanders offer us a great example of Group 3 Humans – those exposed to high-calorie processed foods *and* genetically primed to gain weight.

Reproductive Fitness and the Thrifty Gene

The selection of genes that thrive and survive in times of famine and starvation was first described by the geneticist James Neel in 1962.[13] The phenomenon became known as the thrifty gene hypothesis and it gives us an eloquent explanation for why some ethnic groups suffer with obesity more than others in the same environment.

The thrifty gene hypothesis is based on the theory that people who have an efficient metabolism, or excess fat reserves, can survive periods of famine better than those without. The assumption is that during every famine a certain number of people die, thus thinning out the population, and that the next generation will therefore have hardier genes. The theory offers an explanation for the variations in obesity in different genetic groups, but the mechanism of developing the thrifty gene is actually different to that described by Neel. It seems quite a harsh assumption that swathes of the population were regularly wiped out by famine. There would have been frequent hardship and food shortages, but this stopped short of large numbers of deaths from starvation. A more likely scenario for the development of thrifty genes is that food shortages affected the fertility of the population. If you were a woman with thriftier genes, maybe storing more energy, or fat, than others, you would remain fertile for much longer during a food shortage. Those women without adequate energy reserves would become infertile or lose their baby during pregnancy. The thrifty gene is passed on to the next generation not by physical survival of famine but by the greater fertility of those with more efficient metabolisms in times of hardship. This is known as the reproductive fitness hypothesis.

Don't Cross the Border!

A prime example of the reproductive fitness hypothesis is the Native American Pima tribe. It is thought that this tribe has, over the generations, developed an extreme thrifty gene profile for its population. This has come through many documented, and probably many more undocumented, periods of extreme hardship. A number of the Pima Americans still live in Mexico and lead a healthy, outdoor life, farming and fishing. They have not taken on a Western way of life. These Native Americans do not show any signs of suffering from obesity; despite possessing the thrifty gene they are not exposed to the environmental obesity trigger.

Most of the Pima did not settle in Mexico and now live in the Gila River Indian Community in Arizona, USA. Despite the Pima having their own reservation, much of their traditional way of living has been eroded and replaced by the all-American lifestyle. Their thrifty genes, which would help them survive if America was plunged into a long famine, are unfortunately totally unsuitable for the plentiful, highly processed, hedonic food environment that they find themselves in. Because of the legacy of their past, their thrifty genes now make them the fattest and most unhealthy ethnic group in America. Going back to our cow analogy, they are another example of Group 3 Humans, both genetically selected and environmentally primed to have high weight set-points.

The Pima have obesity rates of 67 per cent, the second highest rate of obesity in an ethnic group in the world, only behind those from the island of Nauru in the Pacific.[14] The rate of diabetes in the Pima is 50 per cent, eight times greater than the US average.

The African Migration

To test our thrifty gene theory further, let us look at another migration that caused terrible attrition rates among settlers in a new land. Enslaved people arriving in America from West Africa had to endure a harrowing transatlantic journey. Packed below deck and manacled in chains, they were treated as sub-human by their captors and were exposed to starvation, beatings, and disease due to poor sanitation. The average duration of a trip across the 'middle passage' of the Atlantic was two agonizing months. In this time, despite only the young and fit being selected to embark on the journey, 20 per cent of the slaves did not survive.*[15]

Here we see another powerful natural selection operating, this time affecting the population of enslaved people arriving in America. The sea passage selected against those with genes that could not withstand famine and energy-sapping diseases such as dysentery. Just as we had seen with our Polynesian sailors, the sea passage favoured those metabolically strong enough or with enough fat stores to survive it.[16] What happened to the population of African Americans, generations later, once they had become exposed to an all-American, Western-type diet? If our theory that obesity is predetermined by inherited genes holds up then they should be at more risk than other ethnic groups living in America (apart from the Pima) whose ancestors had not undergone such attrition. If obesity is not preordained by genetics, then the rates between groups should be similar – because all the ethnic groups in America are equally exposed to Western foods. Look at the statistics below:

Current Obesity Rates in the USA by Ethnic Grouping[17]

All Adults: 35 per cent
Black: 48 per cent
Latino: 43 per cent
White: 33 per cent
The obesity rate among black women living in America is a scary
57 per cent.

---

* It is estimated that, between the sixteenth and nineteenth centuries, approximately 2 million enslaved Africans died during the passage. Another 4 million died in Africa, after capture but before embarkation, on enforced marches and in detention camps. Only 10.5 million survived the passage.

A sad irony for African Americans is that they were enslaved to increase the agricultural workforce, with many working on sugar plantations. The increased availability of sugar as a commodity and its falling price was a by-product of these plantations. Now the new generation of African Americans, still holding the legacy of their metabolically efficient and strong ancestors – genetically primed with a thrifty gene – are again engaged in a fight, this time against obesity and diabetes, brought about by the legacy of the sugar trade.

CASE STUDY – ATOMIC TESTING

'Mr Freeman, please!' my nurse shouted for the next patient to come through to the clinic.

The room darkened for a moment and I looked up from my notes. Mr Freeman's large frame and body had blanketed out the light coming through the door frame. He was the biggest man I had seen, 300kg (47 stone), BMI 90kg/m^2. He was about forty years old, well dressed in elasticated blue corduroys and a home-knitted jumper, softly spoken and intelligent. As part of the consultation I asked him when he had become obese. He told me he had always been big, even as a young child. He had a voracious appetite. The curious part came when I asked him about his family history. 'Who else suffers with obesity in the family?' 'No one,' he replied. He came from a family who were all skinny or normal weight. 'And you are not adopted?' I asked. 'No,' he replied. This staggered me: how could he have reached this tremendous weight without there being a genetic link somewhere. Then he mentioned something as an aside. 'My father worked on the atom bomb during testing.' And we had the reason that he was so different from the rest of his family.

We know that radiation causes increased levels of mutation within a gene. Farmers used to irradiate corn in order to encourage mutations that they wanted to nurture. In Mr Freeman's case, his father had been exposed to radiation during the atomic testing and this had led to a genetic mutation in the DNA that he had passed on to his son: a mutation causing massive obesity.

The Bedouin

Let's go back to the food court in that shopping mall in Dubai. I had observed that the local population, the Emiratis, had a more severe type of obesity problem than other ethnic groups in the mall who were eating the same types of food. We could consider that the explanation for this is the same as the reason that Pacific Islanders, the Pima tribe and African Americans suffer more than other groups with obesity. Maybe the ancestors of the Emiratis had also been exposed to famines that were so extreme that those with 'fat genes' had much more chance of surviving. Maybe they have more of the thrifty gene than other ethnic groups.

I'm not so sure that this is the whole story for the Emiratis. Yes, we know that they originate from nomadic Bedouin tribes. They still proudly carry their legacy of the desert. The two black bands around the white headdress of the traditional male Emirati were used to secure their camels' feet at night to stop them wandering away into the desert. We know life was harsh for their ancestors, but many other ethnic groups have been exposed to generations of hardship and struggle. For example, the ice age which covered northern Europe took hold rapidly and lasted for many generations, and yet the obesity rate for Caucasians of European descent who had endured this hardship is half that of the Gulf Arabs.

Primed for an Oasis Not a Food Court

An alternative theory is emerging as to why the Gulf Arabs suffer so severely with obesity.[18] This theory is gaining acceptance and personally I think it is a more realistic explanation for their current health problems. It has been suggested that the pace of change in their environment is the primary cause for their difficulty in coping with it. Emiratis do not particularly harbour more obese genes than other groups, but their genes have been primed to survive in a harsh environment, without plentiful foods. The theory is based on a new area of scientific research called *epigenetics*.

Previously we had assumed that the genes we inherited from our mother and father were set in stone. It was thought that they could not be altered. This perception is now changing and it has been proved that selected genes can be switched off (in medical terms this is called *methylation* because a methyl molecule covers the gene). The switching off of some genes takes place while we are still growing in the womb; we think that it happens in relation to what the growing baby senses in its environment. The whole process is thought to make the baby more adaptable to the environment that it is born into and therefore more likely to survive and thrive. On the whole this is beneficial to the infant as, in most cases, the environment the mother lives in while the baby is busy developing will be the same as the environment the baby is born into and grows up in. In most cases the anticipated future environment is correct, and the epigenetic moulding of offspring in order to optimize them for the future is usually a force for good. However, as with all predictions, they don't always come true and this is the downside of epigenetics. When a baby is born into an environment quite different from that which was predicted, the baby will struggle to adapt and may develop health problems. I think this may be the case with the Emiratis of the Gulf.

The Dutch Famine Study

Let's look at a famous example of epigenetics when the predicted future environment was wrong. 'The Dutch famine, 1944–1945, and the reproductive process' is a research paper that was published in 1975.[19] It looked at the Dutch famine and how it affected offspring born to the mothers who had lived through it while they were pregnant.

To put the famine into historical perspective, it occurred during a freezing winter towards the end of the Second World War, when the German army was retreating through Holland. During this period, the war was very dynamic with many attacks and counter-attacks; it was a critical win-or-lose time. Because of the nature of the fighting, large areas of Holland were isolated for many months. The harsh winter also froze the canals that would have been used to deliver food to remote areas, thus exacerbating the famine. Strict rationing of available food was introduced, with people allowed to consume only 500kcal per day. The affected areas were relatively large and the famine lasted for six months. The population enduring the famine included young women who were pregnant.

The study was conducted thirty years after the famine and the researchers identified children who had been born to women exposed to the famine. They then compared these offspring to their siblings born before or after the famine. They looked at these two groups and analysed their health as adults. Their findings were surprising. The offspring from starving mothers were, as expected, much smaller than normal when they were born, but once they had reached adulthood they turned out be significantly more obese than their siblings. The type of obesity that the offspring of starving mothers had developed was more dangerous than most – they were more likely to exhibit obesity with fat around the belly rather than on their thighs or buttocks. This type of obesity is seen in men more commonly and it is associated with a higher risk of diabetes and high blood pressure. Not surprisingly, the study also found a higher rate of Type 2 diabetes among the offspring of the starving mothers.

Betting on a Hungry Future

Why would this be? How can starvation in the womb lead to a baby having an increased risk of developing obesity and diabetes in their future lives? Let's look at this from a different angle. What would be the advantage of a baby born to a starving mother having a more robust appetite and gaining weight more easily than normal – or maybe having a more efficient metabolism so that they don't have to burn as much energy as others? The researchers proved that there was no advantage to these traits; in fact they led to a higher likelihood of disease. But what if the organism (the baby) was somehow being clever and could change the way it behaved without changing its DNA (as its DNA is already set)? Imagine it is behaving like a chameleon and changing in response to its environment. What if the starvation environment that the baby had sensed while in the womb was the same environment that he had experienced while growing up – a perpetual famine or food shortage? If this turned out to be the case, then these traits of increased appetite/food-seeking behaviour and a lower metabolism would confer a significant survival advantage to these offspring. This is a classic example of epigenetic priming of genes to predict a harsh future environment. However, in this case, the prediction was wrong: the future environment was not one of famine but of plentiful food. The bet was a losing one – instead of the epigenetic changes providing a health and survival advantage, their legacy was obesity and diabetes.

Another terrible famine occurred during the Biafran War (the Nigerian Civil War), between 1967 and 1970. Researchers looked at over 1,300 babies born before, during and after the war. They had similar findings when they compared the health of the offspring born during the war forty years later. Babies born during the famine were more likely to suffer with central obesity, diabetes and high blood pressure.[20]

The changes in the expression of obesity genes – those that occur in response to famine – are due to epigenetics. This new understanding of genetic adaptation offers fresh perspectives on the interaction between our bodies and our environment. It also poses new questions about evolutionary processes and who we really are.

Can the epigenetic changes that have occurred in one generation of babies be passed on to the next generation? Can the priming of your grandmother's genes when she was developing in the womb – in response to the environment of the time – be passed on to your mother and then on to you? While these questions are still being investigated, there is a suggestion that some epigenetic traits from previous generations do survive for up to four subsequent generations.[21]

Darwin, Lamarck and the Giraffe

When Charles Darwin published *On the Origin of Species* in 1859, the research it contained was ground-breaking, based on his exhaustive work examining and observing animal species and fossils. The subsequent discovery of the structure of DNA by Francis Crick and James Watson in the early 1950s confirmed the mechanisms of evolution, with natural selection and genetic mutation driving it. Darwin's theory is now accepted as the ultimate explanation of our origins. However, recently something has been troubling researchers in this field. They have worked out how long it would take animals and humans to evolve using Darwin's theory and the numbers don't add up. There is not enough time for us to have evolved by simple natural selection or rare genetic mutations. This is where epigenetics may offer us a fascinating alternative theory of evolution that was discredited many years ago.

We know from epigenetic studies that changes to genes, in response to the environment (let's call them 'epi-mutations'), occur 100,000 times more frequently than the simple, old-fashioned genetic mutations of Darwin. Can these epi-mutations affect and drive evolution? There is some evidence that, yes, the gene can be permanently altered by a process called genetic assimilation.[22] If this was the case, then epigenetically driven inheritance would solve the problem of there being insufficient time for Darwin's evolutionary processes to explain our pace of adaptation to the world around us.

The field of epigenetics is new, only a few years old, but it is already challenging traditional theories about how we interact with our environment. Let us spare a thought for a scientist who suggested an epigenetic type of inheritance over 200 years ago.

The Original Theory of Evolution

Jean-Baptiste Lamarck was a French naturalist who proposed an evolutionary theory fifty years before Darwin. His idea was that animals evolved in direct response to their environment, and not, as Darwin later suggested, as a side effect to natural selection. His famous example was that giraffes had been able to evolve a long neck because their direct descendants had spent much of their life stretching their necks to eat leaves and fruit from tall trees.

Because his was the first theory of evolution to be published, Lamarck faced the full onslaught of vicious criticism from the all-powerful Catholic Church for having the temerity to question creationism. His ideas were criticized and discredited by contemporary academics and he was ridiculed in later life when he should have been respected. Unlike Darwin, whom he pre-empted and who became one of the most famous scientists to have lived, Lamarck died in poverty and obscurity in 1829. His theory, like his reputation, is now being resurrected . . . by epigenetics.

A new and developing way of understanding evolution is to accept both Darwin's and Lamarck's theories together. Neo-Darwinism and neo-Lamarckism can underpin our understanding of how we adapt to and evolve to suit our changing environment. With this in mind, let's return to the Dubai food court to reconsider how this impacts on obesity.

Epigenes and the Desert

How can epigenetics explain the Emirati population's problem with obesity? Did the Emiratis' genes make the wrong bet on the future? Let's look at how the environment in the Gulf has changed so rapidly.

Oil was discovered in Abu Dhabi in the 1960s. At this time, the country was made up of several disparate nomadic tribes and pearl-fishing was the mainstay of the economy. By 1970 the UAE was producing 2 million barrels of oil per day (currently it produces 3 million). The founding rulers of both Abu Dhabi and Dubai decided to invest most of the proceeds from oil sales in infrastructure. Massive building projects including housing, hotels, schools, roads and hospitals were started. This 'development' of their traditional Bedouin way of life into a 'Western' lifestyle took only one generation. From enduring baking summers, living in tents, travelling by camel and eating traditional Arabic food, to living in cool air-conditioned apartments, driving Lexus cars and consuming tasty, hedonic, processed foods – all within thirty years.

In America and Europe this transition from the traditional rural way of life to our modern city-based lifestyle took several generations, the changes for each generation being much more gradual in comparison. For Emiratis, the sudden change in their way of life may be what has left their epigenes unprepared. If your genes have been primed to survive in a harsh environment which changes abruptly within a generation to a modern urban one, you will be metabolically unsuited to the environment you were born into. Emiratis probably harbour epigenetic changes that would help them survive the harsh desert environment and nomadic lifestyle, yet leave them metabolically unsuited to the food courts they now encounter. This is the reason that many of them fall into obesity. But what of the next generation?

Eating for Two

We have learned that if your mother was starving or undernourished while she was carrying you, then your genes will be turbocharged (by epigenetic changes) to give you a survival edge. If your environment does not turn out to be as predicted, then your hyper-efficient metabolism is going to give you big problems when you hit the food courts. As a result, your genes will preordain you to become obese.

However, it is not just under-nutrition in the womb that causes genes to change and favour obesity in the Western environment. There is now compelling research showing that maternal *over-nutrition* when pregnant can lead to the development of what is referred to as obesogenic traits in their offspring (*obesogenic* is a relatively new word used to describe things that can cause obesity). Scientists have confirmed this apparent risk in mice. They found that when pregnant mice were over-fed with canteen-type food, their offspring exhibited increased appetite and aggressive food-seeking behaviour, and then became obese, in comparison to offspring of mice that were fed normally.[23]

In humans, increasing levels of blood sugar in pregnant mothers predicts a much greater risk of childhood obesity,[24] and maternal obesity during pregnancy predicts a two- to threefold increase in obesity of offspring by the time they reach four years of age.[25] Intriguingly, when obesity is 'cured' by bariatric surgery (i.e. gastric bypass), the children that the mothers had *after* their weight loss showed no sign of epigenetic transmission of obesity traits, compared to their older brothers and sisters who had been in the womb when the mother was still obese.[26] Dr John Kral, from New York, who co-authored this study, explains that foetuses develop differently during pregnancy depending on their mother's weight and general health – and such changes can be lifelong.

Let's just clarify what we are saying here, because it's not looking good for future generations if they continue to be exposed to an environment that can trigger obesity. First, we found that genes have a 75 per cent influence on a person's weight (remember the study of twins). Now we are saying that if your mother happened to be obese at the time of her pregnancy, she would pass on to you not only her half of your genetic DNA code (which may predispose you to obesity anyway), but also mutations that favour obesity (because your genes were incubated within an obese body).

Why would over-nutrition, or obesity, during pregnancy lead to the development of traits that predispose the offspring to gain further weight? This seems to be counter-intuitive. We can understand why obesity traits are found in the genes of those babies who are predicted to grow in a harsh environment, but what is the advantage in terms of survival of having these traits if the food environment is predicted to be plentiful? The answer may come from the *lack* of micronutrients in the Western diet. Despite a mother being over-weight, she may have vitamin, mineral or other deficiencies because the processed food she is consuming doesn't contain the same nutrition as old-fashioned fresh foods (we discuss this further in chapter 8). The genes in the foetus sense the deficiencies in the mother and the genetic expression of those genes is thus altered to ensure that enough foods are consumed in the future environment to avoid a similar deficiency. Scientists at Duke University have confirmed that administration of vitamin supplements to mice during pregnancy can profoundly alter the way the offspring look.[27]

So there seems to be a U-shaped risk distribution for these traits. In cases where babies' mothers were poorly nourished or had apparent over-nourishment, both groups of offspring pass on epigenetic

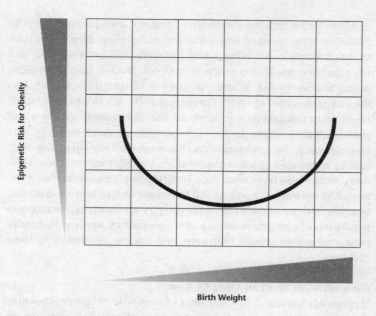

Birth Weight

Figure 2.5 Over-nutrition and under-nutrition of pregnant mothers can lead to a higher risk of offspring developing obesity in adulthood. Studies have described a U-shaped relationship between birth weight of babies and the propensity to develop obesity in adulthood

Source: S. Parlee and O. MacDougald (2014). Maternal nutrition and risk of obesity in offspring: the Trojan horse of developmental plasticity. *Biochim Biophys Acta*, 1842(3), March, 495–506.

traits encouraging the development of obesity in addition to any that are already embedded in their genes.

The School Visit

I remember a visit to an all-girls secondary school with my daughter a few years ago. We were in the process of selecting her next school. What struck me more than the familiar drabness of a London comprehensive was the size of the pupils. The girls who politely showed us around and waited to answer our questions were really suffering – or at least a large proportion of them were – with significant weight or even full-blown obesity.

At the time, I asked myself: how can the proportion of obese children be so strikingly different from my own schooldays? The food that we consumed was quite unhealthy then; how could it be so much worse now? What I had not considered at the time, and what has become clearer since, is the contribution that epigenetics makes to obesity risk. It was not just the environment (which had not changed much) or the children's genes (which would not change much from the previous generation) that drove their weight – epigenetics was a major factor. The obesity in these children was contributed to by epi-mutations that amplified the difficulty they had in maintaining a healthy weight. The mothers of these children, who would have been born from the late 1960s into the 1970s, would have been exposed to the first wave of the obesity epidemic that affected the population in the early 1980s. Many of the mothers would have been obese during their pregnancy and involuntarily promoted epi-mutations that increased the risk of obesity in their children.

Generational Shift in Obesity Risk

If epigenetics really does contribute to obesity in offspring, this may explain the worrying generational shift in obesity risk, with young people being increasingly primed with stronger and stronger obesity traits. Every generation harbours dangerous epi-mutations from obese mothers or grandmothers. Even if our diets stayed the same over the next few generations, obesity will become much more commonplace due to the increasing occurrence of these genetic traits. Remember this when you next see adolescents struggling with obesity. Not only are they having to cope with the normal teenage angst of growing up, but also with much stronger obesity traits than *any* previous generation.

It sounds depressing, but there is a silver lining. If we understand the risk, we can pre-empt it through education. If future mothers are aware that the risk of obesity transmission to their offspring is reversible, they will be much more likely to try and attain a normal weight before pregnancy (hopefully with the help of this book).

The other strategy – for drug companies and scientists – would be to target the obesity genes with their own tailor-made epi-mutations in order to reverse the effect of the gene. In fact, the first groundbreaking study into epigenetics did exactly this. Scientists at Duke University in America were looking at the effect of vitamin supplements on mice, and their offspring, during pregnancy.[28] But the mice they were using were not ordinary mice. The agouti mice had been specially bred to have two traits: obesity and a yellow coat. If a male and female agouti bred, their offspring would always turn out to be just like their parents – large and yellow. When scientists added a simple vitamin supplement into their food, they found that the pregnant agouti mice produced offspring that were brown and . . . skinny. They analysed the genetic code of the offspring and found that the reason the mice had lost their obesity and yellow colouring was that the vitamin supplements had stimulated epi-mutations that had switched off the genes encoding for obesity and coat colour. This research offers us a glimpse into the possibilities of using epigenetics as a treatment for obesity in the future.

There are many different genes that can contribute to people being either obese or lean. One of the first to be identified was the FTO gene. We know people with this gene are on average 3kg (half a stone) heavier than those without the gene. And there are several other genes so far identified that alter the chance someone will be lean or obese. Some of these genes encode for appetite and some for fullness. They determine the amount of food someone will naturally want to eat. They also encode for metabolism, i.e. the amount of energy someone will burn off. We will see later in this book the extent to which metabolism is fundamental to weight control.

If we can eventually target and switch off, using epi-mutations, genes that have been identified as promoting strong appetite or low satiety or low metabolism, we will go some way to fixing the problem of obesity. For now, this is still a long way off.

Keep It in the Family

As far as a genetic predisposition to obesity is concerned, my patients have been telling me about this for years: 'It's in my genes, doc'; 'I'm from a family who suffer with their weight.' Time and time again they will walk in with close blood relatives who are also clearly suffering with their weight. Quite often, once one member of the family has undergone successful weight-loss surgery, other members will follow their lead.

I once did a rare domiciliary visit to a patient's home to assess her for surgery. Her size affected her so much that she couldn't easily make it to the hospital. She weighed 200kg (over 31 stone). I remember the visit because I do not usually see patients in their own homes.

The house was tidy and welcoming, and she had photos of family members scattered around on the mantelpiece, tables and walls. All of them suffered with severe obesity, but were clearly trying to get on with their lives. It really struck me the powerful effect genes can have in predetermining whether you are going to struggle with weight control or not.

CASE STUDY – LIKE ATTRACTS LIKE

A sixteen-year-old Jewish boy came to my clinic with his parents to discuss how best to treat his weight. They were keen for him to get married in the next couple of years, as is the norm in the orthodox Jewish community, but they were concerned his weight might put off future wives. They said they had put him on lots of different diets, but nothing seemed to work. What struck me was the size of the parents. The boy was large, but both his parents were also overweight. The interesting part of this story is that when the parents got married there was no opportunity for bariatric surgery to reset their weight, so their focus was now on helping the younger generation to overcome the obstacles they had been unable to deal with. I could imagine that they, and their families, ended up settling for someone of a similar size in the absence of other suitors. This is an example of what is called assortative mating, where couples attract mates with similar characteristics. In this case that characteristic was their obesity. The unfortunate boy had the triple whammy of receiving obese genes from *both* parents and also obesity-causing epi-mutations as a result of his mother being obese during her pregnancy. In addition to this he lived in an environment with a Western-style diet, and so his obesity genes were triggered. He was a typical example of someone who was almost preordained to become obese.

Summary

So, how does a sacred cow in India explain the cause of the human obesity crisis in the world? Let's recap what we have learned so far.

Farmers can make cows grow bigger by:

1. Feeding them an unnatural special diet (grain/oil mixture)
2. Selectively breeding the bigger cows rather than the smaller cows (unnatural selection).

We found that when it comes to humans we are just like the cows in neighbouring farms. Human populations become more obese:

1. When they are fed a Western diet (grain/oil mixture)
2. When a population has experienced extreme trauma (famines, migrations) that selected only the largest and most metabolically efficient to survive (natural selection, or 'survival of the fattest').

A human population that has survived extreme trauma will become enormously obese if exposed to a Western diet (Pacific Islanders, Pima Americans).

In addition to our genetics and environmental obesity triggers we looked at the new area of epigenetics. This confers an added layer of risk directly onto the genes of people whose mothers experienced famine, or obesity, when they were pregnant. This explains why each generation of our children suffers even more with obesity.

The latest research on this topic seems to validate the explanation 'It's in the genes, doctor', which my patients have been telling me for years. This reassures me that we are now on the right track.

Our stressful, sedentary and sugar-laden lifestyles don't affect everyone in the same way, however. Some people can breeze through life without a thought for their waistline and remain slim; they seem somehow to be protected against obesity, almost as if they are immune to it. Other people go through their lives being stalked at every corner by the spectre of obesity, trying desperately to run away from it (sometimes literally, in the gym), and constantly dieting.

Our genes and our epigenes, triggered by our environment, control our own personal weight set-point. Just like farm animals, most humans have little personal choice as to what size they end up – skinny, slim, average, big or obese. If you happen to have the wrong genes in the wrong environment, then it is almost preordained that you are going to struggle with weight control – it's not your fault. If you try and fight against your weight set-point and consciously attempt to manipulate your weight downwards by dieting, you may, as the next chapter explains, make things even worse. The solution is to create your own personal environment, one that insulates you from those obesity triggers that your genes are looking for.

The final section of this book offers a practical, long-term plan for what to do. But to start things off, if you are struggling with obesity, in my experience the best solution (other than bariatric surgery) is to understand *why* your brain wants a high weight set-point. What signals is it receiving that makes it think it needs extra fat stores? These signals are the key to obesity and how to control it.

THREE

Dieting and the Biggest Losers

Why Our Metabolism Can Change Dramatically

I sometimes watch the reality TV show *The Biggest Loser* with dismay. You are probably familiar with the show's content. The producers select people who are seriously obese and put them through an intensive, thirty-week programme of dieting and physical exercise. The show follows the contestants as they shed the pounds. It sympathizes with the great effort that they are making, focusing on their grimacing faces as they sweat the weight off in the gym. The personal trainer yells threats in their face like a boot-camp sergeant major if they waver in their efforts. But, as the show goes on, we see that all the work that the contestants have put in appears to be worth it.

It's ironic that most of the ads during the breaks in this show are for delicious-looking fast foods and as the programme progresses you get more and more hungry. The show usually ends (while you are devouring a pizza delivery) with smiling contestants being amazed to see exactly how much weight they have actually lost when they stand on the scales. The weight loss can be up to 80kg (over 12 stone), that is, the same weight as an average man! The results seem to be incredible – it's entertaining to watch and pulls in large audiences. But what is the real aim of the show, and all similar boot-camp-type weight-loss shows? Their conclusion is that people really *can* lose huge amounts of weight if they put the effort and application into it. A secondary message is that if you can't do this, you must be weak-willed, greedy or both. These sorts of TV programmes are a major boon for gyms and diet books, but do they really help people who are trying to lose weight?

What *The Biggest Loser* fails to show is the long-term effects on the contestants. We are supposed to believe that the new life the show has given them will be permanent. They are saved, with all the effort they have put in, and have finally beaten obesity.

How does the outcome of *The Biggest Loser* fit in with our weight set-point theory? We can assume that unless the contestants have been able to permanently alter their weight set-point downwards, then the sub-conscious brain would work to bring the weight back up again, using the negative feedback systems controlling appetite and metabolism.*

The Biggest Losers in the Lab

Let's look at a famous study from one of the National Institutes of Health, in Bethesda, Maryland, in the US, conducted by Dr Kevin Hall, a physicist intrigued by the seemingly irregular rules of human metabolism. His team followed up fourteen contestants in *The Biggest Loser* shows and analysed what had happened to their weight and metabolism six years later.[1] The participants had initially lost on average 58kg (9 stone 2lb) each, an amazing result considering how obese they were when they were selected to participate. However, six years after the show, they had *regained* an average of 41kg.

Was their weight set-point still working against them as far as their metabolism was concerned? At the end of the competition their metabolic rate was recorded as being 610kcal lower than when it started. Six years afterwards their metabolisms were even more depressed, at over 700kcal *less* than they were before the show.† This constituted a serious decline in their metabolic rates and meant that just to maintain their weight they had to either consume the equivalent of a large three-course meal *less*, or alternatively go on a 10km run *every day* compared to their pre-diet metabolism. It seems that the contestants' weight set-point was indeed the same as before they were dieting and that their negative feedback systems were doing all they could to win the war and regain the weight that the subconscious brain wanted – despite conscious efforts by the contestants against this.

I Can Lose Weight, But . . .
This would confirm what patients tell us repeatedly about dieting. Yes, it is possible to lose weight in the short term, but in the long term

---

* In this chapter, when I use the word 'metabolism', I am referring to the 'basal metabolic rate', i.e. the amount of energy used in a day before any type of physical activity is added – the amount of energy you would use if you stayed in bed all day (usually 70 per cent of total energy output).

† They were 500kcal less than pre-diet levels after adjustment for weight loss.

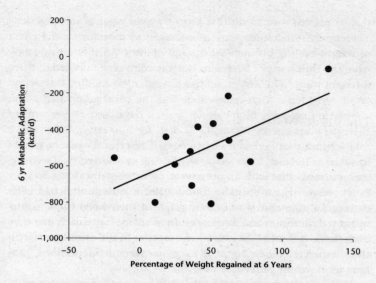

Figure 3.1 Changes in metabolism six years after The Biggest Loser *show*

Note: Contestants who had kept more weight off had much lower metabolisms compared to contestants who had regained weight, demonstrating the weight set-point working to change metabolism years later (statistically significant $r = 0.59$; $p = 0.025$)

Source: E. Fothergill et al. (2016). Persistent metabolic adaptation for 6 years after 'The Biggest Loser' competition. *Obesity (Silver Spring), 24(8)*, August, 1612–19.

they always regain it. It is because the subconscious brain always wins the battle of wills against the conscious brain.

Is dieting detrimental to our long-term metabolic health? If we have been on regular low-calorie diets for many years, how will this affect our metabolism? Will it be lower than when we started dieting? We know from the dietary studies already mentioned that metabolism decreases with weight loss. There is growing evidence that repeated weight loss and then weight regain – so-called *weight-cycling* or *yo-yo-dieting* are detrimental to future weight loss. A study from Korea has shown that people who have frequently dieted lose less fat, and more muscle, when dieting compared with non-weight-cyclers.[2]

Controlled experiments asking people to recurrently diet are impossible to conduct properly – we saw in chapter 1 that in order to have a scientifically supervised diet subjects have had to be confined (i.e. in prison) – so to do this properly over years is not practical. Therefore animal studies are more suited to monitoring the effect of weight-cycling on metabolism and obesity. An interesting study from the University of Bergen in Norway compared mice fed in three different ways.[3] The first group had a regular low-fat diet, the second group were fed a high-calorie diet and the third group had a diet alternating between high-calorie (for ten days) and 70 per cent of their previous energy intake, i.e. a diet (for four days). There was a total of four diet cycles in eighty days. The typical cycle of weight loss during the diet, followed by weight regain when normal feeding, was resumed – but with an overshoot. More weight was put on after every weight-regain episode. If you asked a patient who had been dieting for many years to draw a graph of their weight loss, subsequent weight regain and weight-regain overshoot with each diet they undertook, it would look exactly the same as in the experiment with the dieting mice (see Figure 3.2): yo-yo weight fluctuations, plus long-term weight gain every time.

At the end of the study the mice that had gone through intermittent calorie-restriction diets ended up weighing more than the mice that had been eating a high-calorie diet throughout their lives. Dieting seemed to be counter-productive to weight regulation.

A striking aspect of this study was that the total calories consumed by the dieting mice and those on the high-calorie diet were exactly the same. The dieting mice had somehow developed improved feeding efficiency and a thriftier metabolism, and yet their weight set-point had been nudged upwards by the experience of repeated food restriction.

Why does this happen after diets? Why do we regain the weight that we had lost, then invariably put on even more? I think that each time we diet we are adding to the data that the brain is using to calculate our weight set-point. The brain cannot tell the difference between a diet that we go on from our own free will and a food shortage caused by an environmental catastrophe like a famine. To the brain it is the same thing – both diets and famines equate to calorie restriction and negative energy balance. These incidents are added to the database when calculating how much energy (fat) it would be desirable to store. The more famines/diets the body has had to endure in the past, the higher the subconscious brain will want your weight set-point to be – it wants that insurance just in case the next diet/famine to come along is critical. This fits in with the research out there, plus, more importantly, with the actual experiences of patients

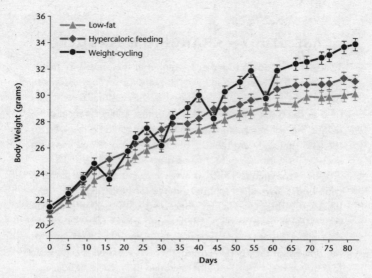

Figure 3.2 Weight gain in mice on a weight-cycling diet compared to those on a high-calorie diet

Source: S. Dankel et al. (2014). Weight-cycling promotes fat gain and altered clock gene expression in adipose tissue in C57BL/6J mice. *Am J Physiol Endocrinol Metab*, 306(2), January, E210–24.

who are struggling with their weight – weight loss, weight regain and then, as the set-point is notched upwards, the weight settling at a higher level than when they started the diet. Recurrent dieting is a great way to train your body to become obese.

Metabolic Variation

In medical school, we were taught that our patients' basal metabolic rates could be calculated if we knew their height, weight, sex and age. Using a complicated equation called the Harris–Benedict formula we would be able to advise patients exactly how much energy they were using and therefore help them to estimate how many calories they needed to consume per day to maintain, or lose, weight. This formula*

---

* BMR (basal metabolic rate): women: BMR = 655 + (4.35 x weight in pounds) + (4.7 x height in inches) – (4.7 x age in years); men: BMR = 66 + (6.23 x weight in pounds) + (12.7 x height in inches) – (6.8 x age in years).

CASE STUDY – CHANGING METABOLISM

Two friends sit down together for dinner at their favourite Italian restaurant. The women had been flatmates for years, cooking and eating together while at university a decade before. Now it was time to catch up. They are strikingly similar in appearance: same height, weight and build, and a bystander would be forgiven for thinking that they are related, but they are not. They are both overweight, but not obese, maybe dress size 12–14.

One of the friends is fretting over the menu: she is starving but can't find a low-calorie option suitable for her; the other friend is not as hungry and is not concerned with calories. As their conversation turns to diets, the starving friend admits to really struggling to hold her weight down. But her old friend reminds her that when they had lived together ten years earlier, they ate together and exercised together and had identical metabolisms.

If we could have checked their metabolic rates ten years earlier we would have confirmed that, yes, they were indeed identical. However, now the starving friend who is looking at the low-calorie option has a metabolism much lower than her friend, probably 200 or 300 calories less per day. The reason? She has been fighting to get from size 14 to size 10 for the last decade – unsuccessfully. This has resulted in her weight set-point being raised to the equivalent of a size 16. The subconscious brain wants size 16, just in case the next diet/famine to come along is more severe. It must protect the body's capacity for survival. Our recurrent dieter is fighting against this losing battle by conscious calorie-counting and denying her appetite, while her body is responding with a lower metabolic rate. We can guess who the winner is going to be.

is the one now used in many smartphone apps to inform people how much basal energy they are burning. These apps are designed to empower the planning of calorie intake in response to metabolic output. Using our first law (energy stored = energy in − energy out), the users can plan their weight-loss strategy. However, there is a fundamental problem with this equation and therefore all apps that use it. The equation calculates the average metabolic rate expected for that person, but it fails to consider the wide variability in metabolisms between people of the same size, shape, age and sex. In other words, the equation ignores our inherent metabolic variability.

10km Run or Three-Course Meal?

If you take a group of ten people who are the same sex and age and size, the Harris–Benedict formula will give you an accurate calculation of the whole group's average resting metabolic rate. If they all had sedentary jobs and didn't go to the gym, then you would expect that they would all use up a similar total amount of energy per day. Let's say in this example the app told us that they used 1,500kcal/day as basal metabolism. However, when you measure each member of the group's actual metabolic rate you would see that there is a striking variability between individuals. The lowest metabolic rate of the group of ten people would be 1,075kcal per day whereas the highest metabolic rate would be 1,790kcal per day.[4] Just as with *The Biggest Loser* contestants after their weight loss, this difference of 715kcal/day is equivalent to the low metabolizer having to go on a 10km run every day in order to have the same energy expenditure as the high metabolizer, or the high metabolizer being able to eat the equivalent of a three-course meal extra every day compared to the low metabolizer!

The difference in metabolism in people of the same size is governed by whether their current weight is above, below or the same as the subconscious brain's desired weight, i.e. their weight set-point. If you are heavier than your brain's desired weight for you, then your metabolism

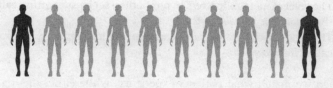

Lowest 10% BMR
1,075kcal/day

Difference = 715kcal/day
Equivalent to 10km run/day

Highest 10% BMR
1,790kcal/day

Figure 3.3 The difference between the highest and the lowest metabolizers in a group of equally sized people

Source: J. Speakerman et al. (2004). The functional significance of individual variation in basal metabolic rate. *Physiol Biochem Zool*, 77(6), November–December, 900–915.

will speed up; if your weight is below your set-point, as it is when you are a few weeks or months into a diet, your metabolism will slow down.

The Dimmer Switch

The average male energy intake is about 2,500kcal per day. This is the same as 10.5 million joules of energy per day. Each day contains 86,400 seconds. From this we can calculate the rate (or power) of energy that an average male uses. The power needed to run a human body is about 120 watts – that's the same power that is required for a light bulb. However, as we have seen, this is just an average. The amount of energy used can range from 60 watts to over 240 watts. Imagine the variability in people's metabolism as being like a dimmer switch attached to a light bulb – it can be set to shine brightly or glow dimly – or anywhere in between.

How Does Metabolism Change?

The dynamic, changing metabolism that I have described is a key feature of our energy-regulating system, but scientists are still uncertain as to exactly how these metabolic changes occur. If they could find the answer to this, then they could target a drug or therapy to halt metabolic changes and make it easier to lose weight on a diet. Having observed hundreds of obese patients, and studied the literature on metabolic variability, I think that the two most likely mechanisms are:

1. The level of metabolic stress in our bodies, set by something called the *autonomic nervous system*.
2. The amount of heat we produce from chemical energy, a process called *thermogenesis*.

When we match what patients experience, while either dieting or over-eating, with the research that is out there, these are compelling explanations of how metabolic changes occur within us.

Run or Fight?
Let's start with the autonomic nervous system. This is called autonomic because it is autonomous or automatic. We have no conscious control over this system. It is also commonly known as the fight or

flight response. The subconscious brain will determine whether we are in a safe environment or whether we are in danger and will adjust the autonomic nervous system (or ANS) accordingly.

Want to know what the fight or flight response looks like in real life? A couple of years ago I was walking in the countryside with my pet spaniel, Maxwell, through a large grass field. As we neared the centre, I noticed that some of the cows in the field, a herd of about ten of them, were going to be blocking our route to the exit gate. I had never had any issue with cows before and would normally have walked straight through them, but something told me to skirt around them this time. In normal circumstances, you would expect the cows to ignore you and go on grazing, but this particular time they had their ears up and then I noticed that they were not actually heifers but adolescent bullocks. As they started their charge towards us, I let go of Maxwell's lead and, for the first time ever, ran like a professional sprinter to the field's 5-ft barbed-wire fence – one I would not normally be able to climb (I'm not very athletic) – but I vaulted over it, landing in nettles with numerous cuts that I could not even feel. As I looked back, I saw the herd chasing after poor Maxwell, ears flapping, who was having a similar autonomic nervous system response. If it hadn't been for our ANS response, both man and dog could have been trampled by these angry young bulls.

Once danger is sensed, we have an innate ability to switch on the afterburners – this is the fight or flight response. Either way, whether we are running away from danger or are cornered and have to fight, we become strong and fast and have more acute vision and clearer thinking. The medical term for this is the *sympathetic nervous system* (SNS) response. Here is what the SNS or fight or flight response does to your body:

1. Increased heart rate and blood pressure to pump blood to the muscles* for running or fighting
2. Sweating to cool the body down during the anticipated exertion
3. Constriction of the blood vessels to the skin so that blood can be preferentially pumped to the heart and brain, leading to a pale appearance

---

* In this chapter I use the word 'muscles' to describe our 'skeletal muscles', i.e. those muscles that are attached to our skeleton and that we consciously use to move ourselves around.

4. Increased blood glucose level to feed the muscles and brain
5. Faster breathing to increase oxygen to the blood
6. Increased oxygen and glucose-rich blood to the brain, increasing speed of thought
7. Dilation of the pupils (for better vision)
8. Release of natural opiates or morphine-like painkillers (called endorphins) in anticipation of injury.

The SNS's fight or flight response is triggered by the hormone adrenaline, which pumps through the bloodstream and activates the SNS, a series of nerves located in the core of our bodies, along our spinal column. It would probably have been an evolutionary survival advantage to have these superhuman traits all the time, but this cannot be the case for one simple reason: energy. The fight or flight response uses up a lot more energy compared to times when we are going about our normal activities. Our SNS reserve kicks in at times of mortal danger.

Time to Relax
The opposite of the SNS, adrenaline-fuelled survival response is the relaxation response. This is due to activation of a parallel system called the *parasympathetic nervous system* (PNS). When this system is more active, our bodies relax into a more energy-conserving state. We are in a safe environment, so it is all right to reduce heart rate and blood pressure; we have more even and relaxed breathing; we reduce blood flow to the brain and just . . . chill.

The traditional way of thinking about the autonomic nervous system is as a way of the body adjusting to different levels of danger – this is what doctors are taught in medical school. But what if this system also had another function? And what if that function were to alter our energy expenditure to offset food excess or deficit? If this were the case, how would our bodies react to energy excess, to over-feeding? We could hypothesize that by activating the SNS, energy expenditure could be increased; just like changing to a lower gear while driving, you don't drive faster but you know you are using up more fuel.

What Happens If You Over-Eat?
How would SNS activation manifest itself in response to over-eating? How would it make us feel if this were really the way that we adapted

metabolically to food excess? We would be more likely to have a high resting pulse rate and to suffer with high blood pressure. We would sweat more than normal; we would have a high blood glucose, which would then stimulate an insulin response (I'll explain this later) and make us crave sweet foods. Our muscles would feel strong. Our brains would be rich in glucose and oxygen and we would feel clear-headed and alive. We would feel good psychologically as a result of the trickle of endorphin painkillers that the SNS provides. Does this feeling sound familiar? Holiday time!

What would happen if this system were also protecting us against weight loss when we went on a diet? In this case, the PNS, the relaxation system, would dominate in order to try and reduce energy expenditure and limit weight loss. Our heart would use up less mechanical energy by reducing our pulse rate (the speed it pumps) and reducing blood pressure (the force it pumps). There would be less blood pumping to our muscles and for this reason they might start to feel fatigued more easily. There would be less blood flow to our brains compared to when we are well nourished, and we would perhaps start to notice that it was more difficult to concentrate; we might even get confused and agitated more easily. We might miss that trickle of wonderful endorphins that we had become accustomed to and therefore feel depressed and empty. Does this sound familiar to anyone out there who has dieted? I firmly believe that what patients describe and what these ANS responses lead to fit very neatly together.

For those of us that do not happen to be on a diet (i.e. most of the population living in our calorie-rich environment, and consuming more calories than we need), what would happen? What effect would metabolic adaptation, in the form of an SNS-type response, have on a population which over-eats? We know from chapter 1 that we are consuming 500kcal more per day than thirty years ago. We also know that most of this excess energy, all but 0.2 per cent of it, is somehow burned off without effort, otherwise we would all weigh over 300kg. A population adapting to over-eating by over-activating their SNS would develop two major health problems: high blood pressure and chronically high blood glucose levels leading to Type 2 diabetes – which are exactly the health problems we see in industrial-ized cities. In addition, the population would find it difficult to wean themselves off the natural opiates, and the feeling of wellbeing, that the metabolic response to over-eating gave them. The food industry might try and profit from this feeling.

SNS Increase – Metabolic Rate Up

There is evidence to back up this theory from Rudy Leibel's research at Rockefeller University.[5] When they performed their research on metabolic changes after 10 per cent weight gain and 10 per cent weight loss they also measured the activity of the subject's autonomic nervous system.

After a 10 per cent weight gain the subject's metabolism was hot – they were burning many more calories, seamlessly and without effort. The researchers noticed that during this time the subject's SNS (fight or flight) activity was increased and PNS (chilling) activity was suppressed. This fitted in with the increase in metabolic rate of 600kcal/day that they measured after the weight gain. The increased SNS activity seemed to be the cause for the high metabolic rate.

When they measured ANS activity after 10 per cent weight loss, mimicking a conventional diet, they found a much more relaxed state, conserving energy by activating the PNS. I would expect that if they had asked the subjects how they felt after the 10 per cent weight loss, muscle fatigue and dulling of thought processes would have been prominent symptoms.

Further studies have confirmed that when humans chronically over-eat their SNS activity is raised, and when they are starved their PNS acts to conserve energy.[6] Curiously, the metabolic adaptation that is explained by this process seems to have been missed by most doctors and scientists. Most labs are just not looking in this direction for their obesity treatment.

So there is compelling evidence that metabolic adaptation takes place and that it is driven by changes in the autonomic nervous system. But there is evidence that there is another way in which our bodies match our metabolism with our food intake and shift the body towards its ideal weight set-point. This second method is called thermogenesis. This theory suggests that extra energy is burned off – literally as heat.

Spontaneous Combustion

Our story of thermogenesis starts during the First World War in a draughty warehouse on the outskirts of Paris. The warehouse was a bomb-making factory and they had just discovered how to make an extra-strong type of dynamite. The lines of workers, mainly women, mixed two chemicals together – *dinitrophenol* (DNP) and picric acid – to make the TNT explosive, before packing it into metre-long artillery shells and soldering them shut. The work was hard and exhausting, but the supervisor noticed that his workforce was not performing as well as expected. The women complained of feeling hot and sweaty, and developing fevers, despite the lack of winter heating in the cold warehouse. After some time, it was clear that many of the women were losing a considerable amount of weight. Then disaster struck. One of the workers, a young woman in her twenties, collapsed with a raging temperature; her muscles were convulsing transiently before they turned rock-hard and stopped working. The paralysis meant she could not breathe. She died, on the factory floor, of asphyxia.

In the 1920s, scientists from Stanford University analysed the effect of dinitrophenol (one of the chemicals used in the bomb factory) on metabolism and found that exposure to the chemical increased the resting metabolic rate by a hefty 50 per cent. Chemical, or food, energy was being converted in the muscle – not to physical energy in the form of movement, but to thermal energy in the form of heat; the side effect was that fat reserves needed to be burned to feed the energy deficit. The heat generated in the muscles raised the temperature of the body and the body compensated by sweating to cool the skin. The chemical dinitrophenol, later to be known as DNP, acted in the core of the muscle cells on the surface of the mitochondria – their cellular engines. These cellular engines normally convert fuel in the form of glucose (from our carbohydrates) to ATP (*adenosine triphosphate*), a molecule which acts like a micro-charged battery that cells can use for building or moving.

Energy (glucose from food) → enters cell → produces ATP (cellular form of energy)

In the presence of DNP, the engines misfire, taking in the glucose but not making ATP, and instead the fuel is converted to heat.[7]

Energy (glucose) → enters cell → DNP blocks ATP production → cell loses energy as HEAT

Miracle Weight-Loss Cure
By the 1930s American pharmaceutical companies had started to produce and market DNP as a revolutionary weight-loss cure. It certainly

seemed to work and within a year 100,000 people had used it. However, the scientists had failed to properly assess the drug's safety and it soon became apparent that it caused several very unpleasant side effects. The first was early formation of cataracts, leading to blindness, and the second was severe hyperthermia (overheating of the body), which led to at least one death. The drug was quickly withdrawn from the market.

DNP made another fleeting appearance in the freezing trenches of the Russian army during the Second World War. Russian scientists modified and weakened DNP and fed it to their troops. It worked – their bodies were miraculously warmed and the rates of hypothermia dropped. The soldiers felt more comfortable, but as their treatment continued they noticed that they were also losing too much weight. Again, the drug was stopped.

More recently DNP has started to make a comeback. Despite its clearly lethal dangers, many bodybuilders continue to use it to lose fat rapidly. It is easy to find and order online. In 2018, four people in the UK died from an overdose and the spontaneous combustion of the muscles that it causes. The final throes of death come when the muscle cells have run out of energy and cannot stop calcium flooding into them, followed by a brief fit and then a rigor mortis-like rigidity of the muscles before death occurs.

The Search for Our Natural Energy-Burner
Knowing the power of DNP to literally burn through our stored (fat) energy, scientists have for decades been striving to find its natural equivalent within our bodies. If a DNP-like fat-burner could be discovered, and somehow safely harnessed, this could lead to a very lucrative weight-loss drug.

Their search began by analysing how 'brown fat' works. Brown fat is abundant in small animals (like mice), which need to keep warm. Unlike white fat (which stores energy), brown fat contains a protein called UTP-1 which, just like DNP, will take food energy and convert it into heat. Unfortunately, adult human bodies do not contain much brown fat, certainly not enough to be able to burn off much excess energy. So recently the search for our natural energy-burner has switched from brown fat to our muscles. Hot-off-the-press research has shown that muscle cells contain a DNP-like substance called sarcolipin, a protein that can, when our bodies so please, burn those excess calories – not by movement or exercise, but simply by converting the calories into heat energy. This heat is then easily lost into the atmosphere. The unwanted excess energy is burned off without effort.

If you are interested in exploring the background to thermogenesis – this fascinating method that our muscles use to burn energy and preserve our weight set-point – in more detail (maybe you are a doctor or scientist), please access the online Appendix at www.whyweeat toomuch.com.

Summary

Let's recap where we are so far with the explanation of the metabolic processes regulating our weight. We have established that our energy reserves, i.e. how much fat our bodies carry, are controlled by our subconscious brain and not our conscious brain. We can try and override our subconscious brain for short periods of time, by dieting, but ultimately our negative feedback processes will draw our weight back to our own personal set-point. The weight set-point is calculated by our brain from our environment, our history and our genes. It can be altered upwards or downwards if we understand the processes involved (this will be discussed further in Part Three). If we over-eat or under-eat, and our weight goes too high or too low compared to our set-point, then our basal metabolic rates are switched upwards or downwards to restore us to our set-point weight.

Metabolism is adjusted like a dimmer switch. If our body wants us to lose weight because it is currently higher than our set-point (for example, after Christmas), it will increase metabolic 'burn'. We have seen compelling evidence that this burn, or metabolic adaptation to over-eating, is controlled via sympathetic nervous system activation (the system we traditionally associate with the fight or flight response). When the sympathetic nervous system becomes more active, we feel the consequences of this. Some of those consequences feel good, like clear thinking and a sense of wellbeing, but others are not so good, like higher blood pressure and glucose levels. In addition to this, SNS activation causes excess energy to be lost by switching on thermogenesis in the muscles. As a result we may feel hot and notice that we sweat easily as the body cools down to compensate for the muscular heat generation.

When our body wants us to gain weight because our weight is currently lower than our set-point (i.e. during a diet), our metabolism can decrease dramatically, down to around 1,000kcal/day. We have seen evidence that this occurs when the parasympathetic nervous system becomes more active. This reduces energy expenditure by the heart (blood pressure becomes normal) and shuts off thermogenesis in the muscles – making us feel cold.

Metabology Rule 1, our first law of thermodynamics (energy stored = energy in – energy out), now seems a lot more dynamic. Energy out varies dramatically – metabolism cannot be controlled by our free will. In the next chapter, we will examine the 'energy-in' part of the equation. Can we consciously control the amount of food, and calories, that we take in over a long period of time, or is this also under some kind of subconscious control?

FOUR

Why We Eat

How Our Appetite (and Satiety) Works

'I'm losing weight but I don't feel hungry any more. Sometimes I have to set my alarm to remind myself to eat lunch.'

This is one of the most common statements that patients make following bariatric surgery. They have tried dieting for most of their lives, but have been unsuccessful every time. They have a perception that they are weak-willed because they always seem to give in to their hunger after they lose some weight on a diet. But suddenly, after bariatric surgery, the veil of guilt is lifted. Their obesity is going away and they feel in control of it. They are losing vast amounts of weight, but are not experiencing the rebound voracious appetite that they have been used to on a diet. Apart from their happiness in being able to lose weight, they are, I sense, relieved to discover that it was not their own greed letting them down. They didn't have the character flaw or weakness that they suspected they had, and which society implied they had, after all. What they had been experiencing with diet after diet were the normal protective hunger signals generated by food restriction. As we saw in chapter 3, in just the same way that metabolism is altered drastically by weight loss, so too are the signals governing how much food that we are directed to eat by our subconscious brain. These 'energy-in' signals are switched off after bariatric surgery.

One of the fantastic consequences of bariatric surgery is that it has stimulated research into appetite regulation. Pharmaceutical companies are well aware of the remarkable changes in appetite that occur after this type of surgery and they want to understand it. Once they have learned about the mechanisms, they can work to produce a drug to mimic the effect of bariatric surgery on appetite – and suddenly they will have a trillion-dollar product on their books. So lots of ongoing research is being funded.

We saw in the previous chapter that our metabolisms are dynamic, varying in order to regulate our weight to the desired set-point. 'Energy out' changes constantly. But what about the 'energy-in' part of our energy-balance equation? How is this regulated?

There are two signals that drive our food intake: the signal to start eating and the signal to stop when we have eaten enough. We know these signals very well:

> ***Appetite***: **producing food-seeking behaviour and the desire for high-calorie foods**
> ***Satiety***: **the feeling of fullness and lack of appeal of food.**

When I was in medical school, our understanding of these appetite and satiety drives – the on/off-switch of energy intake – was fairly basic. We were taught that low blood sugar levels stimulated the desire to start eating and that physical distension of the stomach sent messages to the brain to stop us eating.

With the help of big-pharma-sponsored research, we now know that our appetite and satiety are driven by powerful hormones acting on the brain. Just like our thirst hormones, the satiety and appetite

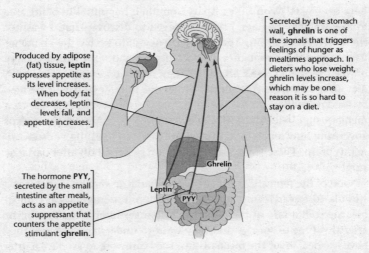

Produced by adipose (fat) tissue, **leptin** suppresses appetite as its level increases. When body fat decreases, leptin levels fall, and appetite increases.

Secreted by the stomach wall, **ghrelin** is one of the signals that triggers feelings of hunger as mealtimes approach. In dieters who lose weight, ghrelin levels increase, which may be one reason it is so hard to stay on a diet.

The hormone **PYY**, secreted by the small intestine after meals, acts as an appetite suppressant that counters the appetite stimulant **ghrelin**.

Ghrelin

Leptin

PYY

Figure 4.1 The appetite and satiety hormones in the gut and fatty tissue

hormones act to change our behaviour without us using our free will to make a conscious decision. As we saw in the Minnesota Starvation Experiment (on pages 21–2), these hormones can literally drive you temporarily insane until the hunger is assuaged.

Our appetite and satiety hormones are produced by the stomach, by the intestines and in the fat tissue (where the energy reserves are sensed). Both organs, the gastrointestinal tract (stomach and intestines) and the fat, are involved in well-regulated negative feedback loops: the hormones travel from the guts or the fat to the brain to ensure that we do not over- or under-eat. These feedback loops can be called the gut–brain pathway and the fat–brain pathway.

The gut–brain signalling pathway controls our short-term, hour-to-hour and day-to-day appetite and satiety regulation. The fat–brain pathway controls our long-term (months and years) energy intake and expenditure.

The Gut–Brain Signalling Pathway

In the 1990s the hormones *ghrelin* and *peptide-YY* (PYY) were discovered in the gastrointestinal tract. Ghrelin is now known to be an appetite accelerator. It is produced in the upper part of the stomach and will increase its levels in response to food deprivation. It usually gets strong enough to tell us to eat at least three times per day. Once we have eaten, the level of ghrelin in our blood drops. Interestingly, it will also stimulate the reward centres of our brain, making food taste so much better when it finally does come along. The longer we go without food, the more we crave it and the tastier it is.

Peptide-YY is produced by the cells of the small intestine in response to food that is inside the small bowel. Once food is sensed to have travelled from the stomach to the intestines, peptide-YY is released into the bloodstream and acts on the brain to produce a feeling of satiety. Not the uncomfortable feeling we get in our stomachs when we have overdone it at the all-you-can-eat-buffet, but the feeling in the brain we get once we have just eaten. There is no desire to seek food any more. The message is quicker and stronger if protein is sensed in the bowel.

What happens to these hormonal appetite and satiety signals when food is restricted, either voluntarily through a calorie-counting diet or involuntarily when there is not enough food in our environment in a famine situation? In 2002, scientists at the University of

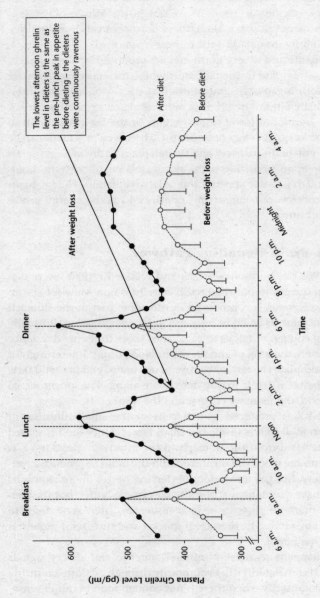

The lowest afternoon ghrelin level in dieters is the same as the pre-lunch peak in appetite before dieting – the dieters were continuously ravenous

After diet

Before diet

After weight loss

Before weight loss

Breakfast

Lunch

Dinner

Noon

Midnight

Time

Plasma Ghrelin Level (pg/ml)

Figure 4.2 The on-switch – appetite hormone levels before and after a diet

Source: D. Cummings et al. (2002). Plasma ghrelin levels after diet-induced weight loss or gastric bypass surgery. *N Eng J Med*, 346(21), May, 1623–30.

Washington measured ghrelin levels in a group of obese volunteers before and after a low-calorie diet.[1] The diet lasted six months and successfully resulted in a 17 per cent average weight loss among the group. Ghrelin levels were measured throughout the day and, as expected, peaked just before breakfast, lunch and dinner. After eating, the ghrelin levels subsided. This pattern – of high ghrelin before a meal and then low levels after eating – continued after the six-month diet had ended. However, the ghrelin signals were 24 per cent higher throughout the day compared to pre-dieting levels. As you can see from the graph, ghrelin levels after dieting were high all the time. In fact, after the diet, the mid-afternoon trough in ghrelin levels – the lowest they got after eating lunch – was similar to the pre-lunch peak that they experienced before the diet. The dieters were literally ravenous throughout the day – even after eating.

This would fit in with how patients describe their appetite after a diet – constantly feeling hungry and having difficulty concentrating on anything but the next meal. When the next meal consists of the low-calorie foods that they have been told to eat, this can be a depressing prospect. The study confirms that appetite drives in dieters are at the very minimum at the pre-lunch levels of non-dieters – but very often much higher.

The Off-Switch – Satiety

What about our off-switch (the satiety hormone peptide-YY)? What happens to this after a diet? And are these signals changed over the long term even when we come off a diet? A separate study looked at a group of patients' ghrelin and peptide-YY levels prior to dieting, immediately after a ten-week diet and then a whole year after the diet had finished.[2] The results are depressing for anyone who has tried voluntarily to lose weight by calorie-restricting – but they do explain how dieters have been feeling. This study found that after the diet, ghrelin levels – and therefore appetite – were elevated, just as had been seen in the previous study. In addition, the satiety signal provided to the brain by the hormone peptide-YY was significantly lower. So the dieters were hungrier and when they ate they experienced more reduced feelings of satiety than they had before the diet. Sort of expected, but now we have the bad news.

One year after the diets had finished, when the group had regained most of their weight, the levels of ghrelin (and therefore the appetite level) remained higher and the levels of peptide-YY (the satiety feeling) remained lower than pre-diet levels. Not only had the diets not worked to maintain weight loss, but the dieters' appetite- and satiety-signalling remained disrupted a whole year after stopping their diet. Life had just got even more difficult for our group.

Again, the results of this study match exactly what patients themselves describe after low-calorie dieting. Many express the feeling that their problems with weight regulation really started when their doctor or dietician (or, on many occasions, the school nurse) told them to try and consciously lose weight by low-calorie dieting. We will talk about diets in more detail in chapter 12.

The conclusion? We already know diets don't work in the long term. What is emerging, though, is that diets can become counter-productive and can actually stimulate longer-term weight gain. The only way to lose weight is to understand what controls your metabolic and appetite drives – once you have this knowledge you can use it to adjust your weight to a healthier and more stable long-term level. Part Three of this book will guide you through these processes, but first it's important to understand how your body-weight regulation works – only then will the changes set out in Part Three become a permanent part of your life.

The Fat–Brain Signalling Pathway

Our fat cells are in direct communication with our subconscious brain via a messenger hormone called leptin. This hormone is the powerful master regulator of our long-term energy stores – it works over weeks and months, rather than hours and days like the gut hormones. It controls both the long-term appetite and satiety drives (energy in) as well as the metabolic rate (energy out). Leptin is released by our fat cells and the amount of the hormone in the circulation mirrors the amount of fat that we have available as our energy reserve.

The leptin messenger tells the *weight-control centre* of our brain the status of our current nutrition. This is the simple, but very powerful, fat–brain signalling pathway. It is a bit like the petrol gauge of our car, which shows us how full the tank is. Leptin levels are high when we are carrying a lot of fat and low when we are slim. If fat reserves are depleted, leptin will direct the brain to become hungry and to eat – to take in energy and preserve what we have. If fat reserves are plentiful, leptin will take hunger away and direct the body towards reproduction, or growth and repair.

Leptin levels tell us whether to go out looking for food or to go out looking for a mate. Basically, leptin enables your fat stores to speak to your brain, letting it know how much energy we have stored and, most importantly, what to do with this energy.

The word 'leptin' derives from the Greek *leptos*, meaning 'thin'. Leptin, when it is working properly, will do just this – make you thinner. When the fat–brain axis is functioning correctly, a person will be able to relatively easily maintain a stable body weight over a long period of time without any type of conscious control of calorie intake or any need for extra energy expenditure in the gym. All this is down to leptin – our powerful metabolic thermostat. By directing energy in as well as energy out, leptin exerts long-term control over our energy reserves. The leptin signal means that our energy stores are able to self-regulate in a classic biological negative feedback loop.

When leptin is high, food is off the brain's agenda and we are free to daydream about other things. In addition, via stimulation of the sympathetic nervous system, leptin increases our metabolism, meaning that seamlessly, without any effort, we burn off excess energy without even having to get out of our chair.[3] When leptin is working, it is a wonderful hormone to get our weight back down to where the subconscious brain wants it to be – down to its set-point.

Some people, who can keep their weight stable for years, may congratulate themselves that they are able to consciously control their holiday weight gain of a few pounds by sweating it off at the gym after their return, and maybe also by calorie-counting for a while. But, in actual fact, leptin is the boss. High post-holiday leptin levels, caused both by weight gain and by over-eating, will lead to a much higher level of metabolic energy expenditure per day than any half-hour jog, effortlessly. The post-holiday diet to get the excess pounds off is aided by leptin cutting off appetite and food desire. The diet seems easy and the exercise works better than expected: normal weight is regained (Figure 4.3). If you fight the battle with leptin on your side you will win easily, but even without conscious effort your weight would have eventually settled down towards your original set-point anyway; it would just have taken slightly longer.

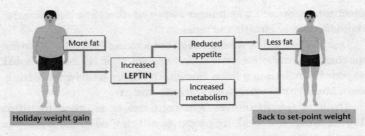

Figure 4.3 How the action of leptin aids a reduction to the weight set-point

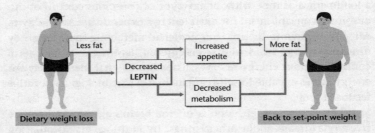

Figure 4.4 How leptin can act to cancel out weight loss through dieting

In a similar way to excess holiday weight coming off more easily than expected, thanks to our leptin thermostat, we can also explain the difficulty of continued weight loss on a low-calorie diet. Remember, leptin is the master controller of our energy stores. If these stores differ from what our subconscious brain perceives to be the safest weight, i.e. our own personal weight set-point, leptin will act to correct the difference. If our weight falls below our weight set-point (this is usually because we are trying to lose weight by consciously dieting), then leptin levels fall, as fat is lost. The effect of this? Plummeting metabolic rate and a voracious appetite. We may win a short-term battle, but leptin will always win the war between our conscious and subconscious will to eventually return the body to the desired weight set-point (Figure 4.4).

Leptin was discovered in 1994 by researchers at the Howard Hughes Medical Institute, Rockefeller University, New York. The scientific team, led by Jeffrey M. Friedman, discovered a way to breed mice that were unable to make leptin.[4] As the mice lacked the leptin-making gene, their fat cells could not manufacture it. They compared these mice to normal mice and found that the leptin-deficient ones developed a voracious appetite and exhibited massive weight gain. Despite being fed the same types of food, the mice which had no leptin in their bloodstream were soon double the size of their neighbours. Even when they were clearly struggling with severe obesity, these mice still exhibited the behaviour of a ravenously starving animal. Although they now had enormous fat reserves, their fat tissue was unable to produce leptin, which led their brains to assume the 'petrol tank gauge' read zero even though the 'tank' was overflowing.

When the researchers injected the leptin-deficient mice with a leptin replacement, their behaviour suddenly changed. They no longer ate voraciously and seemed to have more energy. After a series of leptin injections, the rats eventually lost all the excess fat that they had gained – and their obesity was cured.

Researchers at the University of Cambridge were the first to discover a similar genetic deficiency of leptin in humans. In 1997 Dr Sadaf Farooqi and her team in the Metabolic Disease Unit investigated two cousins of Pakistan origin who had extreme early-onset obesity.[5] The cousins, both girls, were aged eight and two. They'd had normal birth weights but both had exhibited a constant ravenous hunger thereafter. If food was denied them, they would develop severe behavioural disturbances with tantrums and violent mood swings. The elder cousin had already undergone liposuction before the age of eight, but to no effect, she weighed 86kg (13 stone 7lb). Her two-year-old cousin already tipped the scales at 29kg (4 stone 7lb). When the team of researchers tested their leptin levels they found that despite their having massively excessive weight, and fat reserves, they had almost no leptin in their bloodstream. The signal to the rest of the body that their fat stores were excessive was absent. In fact, the opposite was the case: the extremely low levels of leptin were a strong signal to the body that energy stores were critically low. The cousins' aggressive voracious eating behaviour was a normal response to this perceived mortal danger of starvation.

The next step for the Cambridge researchers was to mirror the successful treatment of leptin deficiency that had been reported in the original study of genetically leptin-deficient mice. The cousins began a series of leptin-replacement injections. Immediately, just as in the animal studies, their behaviour changed, their appetite decreased and they started to lose considerable amounts of weight.

These were exciting times for obesity research scientists around the world. There was hope that finally, after years of effort, the definitive cure for obesity had been found. It was assumed that leptin, when injected into obese people, would cure them of the condition. Pharmaceutical companies and their chief researcher scientists jostled for position to be able to own this product. Massive amounts of money were at stake. This would be the trillion-dollar drug they had been looking for.

But then the results of the scientific studies started to be published. Several different research groups tried, and failed, to produce weight loss in obese humans by injecting them with leptin.[6] They measured the leptin levels of their obese subjects and found they were elevated, and yet the leptin signal did not seem to be getting through to the appetite- and metabolism-controlling centres in the brain. In fact, when compared to a placebo treatment (water instead of leptin), there was no difference in weight loss.

What was the difference between the successful treatment of genetic obesity in the young cousins – and also in the leptin-deficient mice experiments – compared to the failure of treatment in subsequent human trials? When the researchers looked at the leptin profiles of a normal obese person in the human trials they found that their leptin levels were high, reflecting their level of obesity. These people had gradually put on weight throughout their lives, compared to the Pakistani cousins who had voraciously eaten since birth and put on weight rapidly. The cousins had very low, almost zero levels of leptin, while the obese adults tended to have high levels of leptin. It soon became apparent that inherited genetic disorders causing low leptin levels were extremely rare. In fact, after the discovery of the leptin-deficient cousins, only fifteen similar cases in the whole world have subsequently been identified. The genetic mutation to cause leptin deficiency needs to be transmitted from both the mother's and father's genes. As these mutations are extremely rare, they tend only to appear in consanguineous relationships where close families intermarry.

The next question we need to examine is how could someone become obese in the presence of high levels of leptin? This seems to be the norm in most people suffering with severe obesity. What has gone wrong with leptin, our powerful fat controller? The next chapter explains why the master regulator of our weight can stop working.

Summary

We have learned in this chapter that our appetite (that uncontrollable urge to eat) and our feeling of satiety (that feeling that sufficient food has been taken in) are strongly controlled by newly discovered hormones that originate in our stomach and our intestines. The stomach hormone, ghrelin, tells us to go out and seek food – it's the signal to take in energy in the form of food. The hormone peptide-YY, originating in the intestines, sends us the message to stop eating – we know we have had enough food for now.

These hormonal signals are extremely powerful: appetite can be like a parching thirst and strong feelings of satiety can make you feel nauseous. They are designed to be part of that negative feedback loop – trying to keep our weight at the brain's perceived safe weight set-point. Lose too much and you will be ravenous and never feel full. Gain too much and you will lose your strong appetite and feel satisfied without food. In the fight for your weight your subconscious brain will always win – forcing you to take in the energy that it wants.

When combined with the dramatic decrease in our metabolism (energy expenditure) that can occur to stop weight loss (as discussed in the last chapter), it should be becoming clear that the old-school weight-loss equation – energy in (food) – energy out (metabolism) = energy stored (fat) – is not under our conscious control.

We learned that bariatric surgery works by dramatically altering the appetite and satiety signals from the stomach and small intestine to produce seemingly effortless weight loss. People who have undergone this type of surgery are relieved that their appetite was not really part of some sort of character flaw – that it was not in fact under their control at all.

Finally, we learned that leptin, the hormone produced by our fat cells, is the master controller of our weight. It works to stop us getting too fat by telling the brain how much energy is already stored – by acting in the same way as the petrol gauge on a car. Too much leptin means little appetite and a high metabolism – producing weight adjustment back down to the set-point. It helps direct both metabolism and appetite/satiety to keep our energy reserves on an even keel, preventing runaway weight gain and weight loss. In many people leptin is the reason that their weight is not an issue. They don't count the calories as leptin is in charge. When this hormone is deficient, as occurs in a very rare genetic condition, staggering and rapid weight gain occurs.

But if leptin is really the master controller of our weight, then why do people who suffer with obesity have such high levels seeping into their bloodstream from their fat? Why does leptin not seem to be working for them?

The next chapter explains why the master regulator of our weight can stop working.

FIVE

The Glutton

Understanding the Fatness Hormone

As I sat down for breakfast, I took a moment to look around me. I was in Dubai for my clinics and the sun burned down on the terrace outside – the city and the Burj Khalifa tower shimmering in the distance. The murmur of morning conversations and the chink of cutlery filled the hotel's dining area, with couples and families and singles enjoying the sumptuous buffet. I poured my tea, but was interrupted when I sensed a momentary silence in the atmosphere. I looked up and paused – the diners were gazing at a gigantic man who'd just entered. He was dressed in a traditional white Arab *kandura* robe, this one presumably specially made for him as it was as wide as it was long. He wore no headdress and I guessed from his hairline and the scattered grey in his stubble that he was in his forties. He moved well in spite of his massive size and round shape, but as he sat down at a table opposite me I noticed desperation in his eyes. He looked pale and was sweating profusely despite the cooling air-con. He was not a tall man but must have weighed 200kg (over 31 stone). He was breathless, though trying to hide it – and he looked as if he was really suffering.

I covertly observed this poor man for the next hour as I ordered refills of tea. His behaviour was remarkable. He went to every station in the buffet and ordered the waiter to take each plate that he filled to his table. Eggs, hash browns, chicken sausages and beans piled on one plate, cold meats and cheeses on another; a large bowl of fruit cocktail was filled to the brim; Arabic flat bread and hummus, toast and jam; two plates piled high with cakes and croissants; and three large glasses of fruit juice. When he finally sat down ready to start breakfast, his table (which would normally seat four) was totally covered with plates of food, enough to feed ten people for breakfast. He ate extremely efficiently and very fast, but there was still desperation in his eyes. Within twenty minutes he had consumed the entire table of food. He ushered the waiter over for more . . .

When the man had finally finished, he dusted himself down, raised himself out of his chair and walked confidently out. He looked much better in himself, his colour had returned and the haunted look had left his eyes. But I was left wondering how could one man eat so much and so fast? Was he greedy or had he become somehow addicted to food? Or was there something else going on – was his gluttony a symptom of an underlying disease?

As for my fellow diners, who muttered comments to each other as he left, their views were obvious. Their subtle, out-of-his-view shakes of the head and empathetic glances towards each other told me their verdict. This man was guilty as charged. He was fat because he ate too much and he ate too much because he was a glutton. He had committed one of the seven deadly sins in full public view and showed no remorse.

But what if reality were different? Let's look at things from the man's perspective. If we had asked him how he felt that morning, he may have told us that he had had a fitful night's sleep, probably waking every hour as his body struggled to get oxygen into his system through his loud snores. Because of the low oxygen levels in his brain he would have described waking with a headache. As he got ready for the day, he would have been aware of a constant level of stress and anxiety on account of the way he looked and the way other people perceived him. But the main thing he would have remembered about that morning, apart from the headache and anxiety, would have been his ravenous hunger, the feeling that he had not eaten for a week, despite the fact that he binged on food every day. That's why he looked so anxious and pale when he entered the dining area. Maybe hunger signals were driving his behaviour?

But surely this man would have had an excessive amount of leptin in his bloodstream? Leptin levels should rise with increasing levels of fat. This should have had the effect of *decreasing* his appetite and *increasing* his metabolism. So, what had gone wrong with the feedback mechanism that was supposed to stop this poor man getting so obese?

If we had measured his leptin levels, we would have been able to confirm that they were appropriate to the amount of fat that he was carrying: they were sky-high. So why is leptin, the hormone that controls fat storage month by month and year by year, not working?

The answer to this leads us to the root cause of the disease of obesity. The clue comes from the studies of obese human volunteers who failed to lose weight when injected with leptin. In the experiment, leptin levels were high prior to the injections. Increasing an already high leptin level therefore had zero effect. Leptin did not seem to be working any more.

In the Pakistani cousins who had a rare genetic leptin deficiency (see p. 87), the injections had a dramatic effect, enabling them to lose considerable amounts of weight. Leptin seemed to work appropriately when the levels in the body were low, but when the levels were high, it stopped working.

The scientists concluded that at high levels leptin's message to the brain starts to get disrupted. High levels of leptin are present, but the brain can't sense this. When leptin reaches this threshold, a condition called *leptin resistance* develops. The brain is 'blind' to the high leptin levels and therefore to the high fat reserves. In fact, the opposite message is getting through: the brain senses a much lower leptin level than is actually present and interprets this as starvation status. As we observed at the Dubai breakfast buffet, this can lead to increasing hunger and a voracious desire to stave off starvation. The effect? More weight gain, more fat-producing leptin, even higher leptin levels and even more leptin resistance. The fatter the man becomes, the hungrier he feels; the more he binges on food, the fatter he becomes. This vicious cycle of increasing weight gain and increasing leptin resistance, leading to more weight gain, describes end-stage, fully fledged obesity.

Let's go back to our petrol tank analogy. Imagine that you are driving your car and you notice that the fuel gauge is dangerously low. You immediately start to worry about finding the next petrol station. You need to fill up as soon as possible – it's urgent. Little do you know that in fact you have a full tank of petrol. The problem is the fuel gauge – it's broken. This is the same as leptin resistance – the brain thinks there is no fuel (fat) on board, but in fact there are plentiful reserves.

The Tipping Point – Leptin Resistance

The holy grail for obesity research is understanding, and then fixing, leptin resistance. If this can indeed be reversed, if the brain is able to recognize the high levels of leptin present, then it could correct itself. Triggering this change would mean sufferers' voracious appetites and low metabolisms could be reversed and their weight would normalize – to continue the analogy, the broken fuel gauge would be fixed and there would be no need for unnecessary emergency stops at the petrol station.

A clue to leptin resistance comes from our emerging understanding that it not only controls the amount of energy we store, but also tells us what to do with those energy reserves. Our DNA wants two things from us: survival and reproduction. Once we have reached adulthood our reproductive success is dependent on our nutritional status. If a young woman is not carrying enough fat, or energy reserves, then there is a risk that a pregnancy may not progress if food becomes scarce. If there are abundant fat reserves, then it is much more likely that a pregnancy will be successful, even if a food shortage has occurred. It therefore makes evolutionary sense that leptin, the fat messenger to the brain, only stimulates reproductive behaviour at times when it is nutritionally appropriate. This has in fact been confirmed in the research.[1] Leptin, acting via an intermediary messenger, stimulates gonadotropin-releasing hormone (GnRH), which goes on to tell the ovaries to start functioning. An interesting side effect of the obesity that I see in many patients is a condition called polycystic ovary syndrome (PCOS), in which the ovaries stop functioning normally and the patient becomes less fertile. It may be that in the future we find that leptin resistance is contributing to this condition.

We also know that, in people suffering with severe weight loss through famine or illness, infertility rapidly ensues in order to protect the body against a dangerous and energy-sapping pregnancy. So leptin not only acts as the messenger to the metabolic and appetite areas of the brain, but also acts to switch on or off our reproductive ability, depending on our nutritional status.

Leptin Resistance Can be Entirely Normal

Leptin resistance causes weight gain by stimulating an inappropriate appetite when there are already ample energy reserves. It basically stimulates lots of energy in and gives us room to grow. There are two periods of our lives when leptin resistance can benefit us by helping to stimulate this growth:

1. Pregnancy[2]
2. Adolescence.[3]

Healthy leptin resistance at these times is crucial to our survival – without growth and reproduction we would be extinct. But remember, in both healthy leptin resistance and obesity-related leptin resistance, the signs are the same – ravenous hunger (to take in energy) and fatigue (to preserve energy). We forgive our teenagers, and our mums-to-be, for this behaviour, but spare a thought for our hungry and tired sufferers of obesity – they are getting the same signals.

What Causes Leptin Resistance?

There are many theories in the scientific literature as to the cause of leptin resistance,[4] and it is still being debated. But in my opinion the likely causes are a combination of:

1. The hormone that controls the glucose level in our bloodstream – *insulin**
2. A protein that controls inflammation in the body called *TNF-alpha*, that triggers:
   o inflammation of the weight-control centre of the brain
   o the need for even more insulin.

Insulin and Leptin

We have learned that leptin is the long-term controller of our weight. When our fat stores increase, the level of leptin in our blood also rises. The hypothalamus, which controls our appetite and metabolism, uses the leptin level as a guide to how much fat we are carrying and will adjust appetite (how much energy we take in) and metabolism (how much energy we use up) according to the levels of leptin (and therefore the fat reserves). Leptin works by attaching to special cell receptors in the hypothalamus. The receptors act like a cellular letterbox where the message – too much fat – can be delivered. However, the strength of the leptin signal to the hypothalamus can be diluted by the hormone insulin.

Both leptin and insulin send a signal to the same cells of the hypothalamus. The cells have separate receptors (cellular letterboxes) for either leptin or insulin. However, once the message is delivered,

---

* Insulin is a hormone released by the pancreas when we eat any food containing glucose (e.g. sugar, bread, pasta). Its job is to transport the glucose from the blood into the cells themselves so that they can use it for energy.

the signal pathways within the cell overlap each other. The cell cannot read both the insulin and leptin messages at the same time. So if insulin is acting on the cell receptor, then there is no room within the cell to read the leptin signal as well, even if leptin is present and posting its message. The leptin message therefore goes unread.[5] As a result, the hypothalamus thinks fat stores are low (when they are excessive) and stimulates appetite while at the same time cutting metabolic energy expenditure. Just like the faulty fuel gauge in your car showing empty when the tank is actually full.

High insulin → leptin resistance

The profound effect that insulin has on leptin resistance means it too becomes critical in the control of our weight set-point. Higher levels of insulin mean more leptin resistance and more leptin resistance means a higher weight set-point – and therefore a higher weight. We will discuss insulin in much more detail in chapter 10.

TNF-Alpha

TNF-alpha is released by cells that act as police against infection or injury. These cells (called macrophages) roam around our bodies looking for potential trouble (e.g. damaged cells or invading bacteria/ viruses). Once trouble is spotted, the cellular police release TNF-alpha (think of police officers with pepper spray or a Taser) and this stimulates a cascade of events leading to inflammation (representing the arrest and disposal of the threat and repair of the damage). This is part of the normal inflammatory response.* However, in obesity, once fat cells reach a critical size, the cellular police are called into action to investigate.† They assume the swollen cell is damaged and therefore release TNF-alpha to start the repair process – but this can have untoward side effects.

Obesity → swollen fat cells → increased TNF-alpha → inflammation

---

* Inflammation is essential for our health – it fights infection and repairs cells. Signals from damaged cells and from foreign invaders into our bodies stimulate our inflammatory response.

† There is some evidence that the leptin secreted by the fat cell may also be responsible for the recruitment of cellular police, but the result is the same: inflammation around the fat cells and inflammation spreading throughout the body.

The chronic reaction of our bodies' 'police' against swollen fat cells means that higher than normal levels of inflammation are always present. Obesity is a pro-inflammatory condition. This fits in with patients that I see in my clinic who suffer with obesity and are enquiring about bariatric surgery: all of these patients will have a positive blood test for inflammation (called a CRP test). Below I explain why this is critical to understanding obesity.

TNF-alpha is also increased in response to a typical Western diet that has a low omega-3 to omega-6 ratio. We will discuss this in more detail in chapter 9.

Western diet → increased TNF-alpha → inflammation

Inflammation of the Weight-Control Centre
We know obesity (and the Western diet) causes the production of TNF-alpha, which stimulates further inflammation throughout the body. All organs of the body are affected to some degree. There is a higher rate of inflammation in the blood vessels (causing heart disease), in the joints (causing pain and arthritis) and in the cells (increasing the risk of many cancers).

But there is now emerging evidence that the inflammatory reaction causing obesity in the body also has a direct effect on the hypothalamus – yes, your weight-control centre, the area that depends on the leptin signal to calculate your weight set-point. And the result of hypothalamic inflammation is leptin resistance.[6] The fat signal is not getting through – high levels of leptin are not being sensed – meaning that you perceive famine in times of excess.

Obesity-driven inflammation → hypothalamic inflammation → leptin resistance

From an evolutionary perspective, it makes sense that if we are sick or seriously injured the inflammatory response that is initiated also causes leptin resistance. Any injury will take energy to heal it, so an appropriate response would be to consume more energy than would normally be required (taking into account our current fat stores). This is achieved by blocking the effect of leptin, thus leading to increased hunger and subsequently more energy intake (in the form of food).

TNF-Alpha Impairs Insulin Strength

The other effect that TNF-alpha has on leptin is via its influence on insulin. When the level of TNF-alpha in the blood increases (as it does with obesity-related inflammation), it acts to block the effectiveness of insulin.[*] Insulin becomes inefficient at doing its job of transporting glucose into cells (this is called insulin resistance[†] in medical language). The result? More insulin is produced by the pancreas to compensate.

> **Obesity-related inflammation → increased TNF-alpha →**
> **decreased insulin efficiency → increased insulin →**
> **leptin resistance**

An example of a healthy *high* TNF-alpha level is in pregnancy. TNF-alpha is produced by the placenta as the pregnancy progresses. It has a critical role in modulating the body's immune response to the growing foetus.[8] If immunity is not altered during pregnancy the foetus will be recognized as foreign and this will elicit an immune reaction against the baby that would end the pregnancy. As TNF-alpha levels increase in pregnancy, so does its effect on blunting the effectiveness of insulin. Diabetes in pregnancy (also called gestational diabetes) is common and increasingly researchers are recognizing the role of TNF-alpha as a causative factor in its development. As well as its beneficial effects on immunity in pregnancy, high TNF-alpha will lead to leptin resistance and stimulate the appropriate energy intake (and weight gain) of pregnancy.

The effects of inflammation on leptin-signalling in the hypothalamus are beneficial in helping to balance energy reserves at times of injury, and helping the positive energy balance required for the growth of a pregnancy. But they become detrimental in obesity, driving further weight gain and increasing the risk of diabetes and heart disease.

Now let's return to the man at the Dubai breakfast buffet and see if we can explain his behaviour from what we have learned about leptin resistance. This man had very high leptin levels, in accordance with the amount of fat he was carrying. However, his large fat cells

---

* By down-regulating tyrosine kinase activity in the insulin receptor.
† Insulin resistance and obesity go hand in hand – 90 per cent of incidences of Type 2 diabetes (caused by insulin resistance) are overweight or obese.

(1) caused a chronic inflammatory reaction and (2) led to high levels of TNF-alpha.

- The inflammatory reaction in his body attacks his set-point directly by producing leptin resistance in his brain.
- The TNF-alpha acts indirectly to produce leptin resistance by causing high insulin.
- The high insulin (stimulated by a Western diet and TNF-alpha) results in a blockage in the leptin-signalling in the brain.

Finally, it is likely that this man has full-blown Type 2 diabetes, leading to even higher levels of insulin and even more pronounced leptin resistance. The result? An extremely obese man in a vicious cycle of voracious hunger and weight gain as a result of leptin resistance. It is almost as if his fat is acting like a tumour – encouraging itself to grow inexorably by sending the wrong metabolic signals to his body.

Leptin Resistance Can be Reversed
We know from animal studies that leptin resistance can be artificially reversed. Rats who are fed Western-type foods (high in sugar and oil) respond by developing insulin resistance – which leads to leptin resistance and then weight gain. When the animals are put back onto their normal diet, their leptin resistance and insulin levels stabilize and their weight returns to normal levels.[9]

Higher insulin levels mean more leptin resistance. Unfortunately,

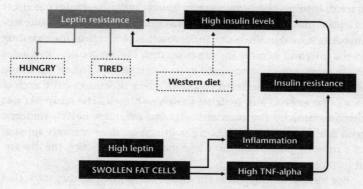

Figure 5.1 The vicious cycle leading to leptin resistance

our current Western diet predisposes us to high insulin levels. If we consume a meal or snack that contains a lot of sugar or processed carbohydrates like wheat, this will result in a spike in the amount of insulin produced (to process the glucose into our cells), and in the West we are surrounded by this type of food. (This will be discussed further in Part Two.)

Summary

We learned in chapter 4 that leptin, the hormone produced by our fat cells, is constantly working to try and keep our weight on an even keel. When we over-eat, and lay down too much fat, our leptin level increases. This is sensed by the weight-control area in our brain (the hypothalamus) and leads to powerful unconscious changes in our behaviour. Hormones work to decrease our appetite and increase our metabolic rate. Our food intake is lowered and energy expenditure increased – thus regulating weight gain. This is how most people, with only moderate effort, can maintain a regular weight for months and years.

We learned in the present chapter what happens when leptin stops working. The message that we have enough fat – enough energy in reserve – does not get through. Like a faulty fuel gauge in your car, where, even though the tank is full, the gauge shows empty, you feel the urgent need to fill up.

When there is a combination of high insulin and lots of inflammation in the body, leptin will stop working. The leptin → insulin → inflammation interaction becomes a downward spiral once it gets started. Insulin (caused by too much sugar initially) dilutes the effect of leptin. Inflammation (caused by a Western diet) stops the dilute leptin being sensed in the brain. Inflammation blocks insulin-signalling in the cells and so more insulin is needed. Even more insulin dilutes leptin even more . . . and so the cycle continues, leading to worsening leptin resistance. The brain perceives a lack of leptin (and therefore lack of fat reserve) and switches to survival mode. The result? As was demonstrated by the man at the Dubai breakfast buffet, someone who is already excessively obese is driven by the voracious appetite of a starving man to eat more and more. This is obesity, the disease, at its most extreme.

But we also found light at the end of the tunnel. Studies show that leptin resistance can be reversed by changing the quality of the food that is consumed. This solution forms the basis of Part Three: how to lose excess weight and then maintain a healthier weight.

Metabology

We have now completed our short course in metabology: how the body deals with energy regulation, over-consumption of food and dieting. The crucial elements of metabology are the weight set-point theory and its effect on the vigorous defence of a fat store that is calculated by the subconscious brain to be appropriate and healthy (considering our genes, our current environment and our past history). Finally, we have seen what happens when a threshold of obesity is reached that leads to leptin resistance and the vicious cycle of obesity driving hunger and hunger driving obesity.

SIX

The Last Resort

How Weight-Loss Surgery Works

Do I shock you? We are very playful here. It's a good tone for an operating theater. It is a theater, after all.

David Cronenberg, *Consumed* (2014)

The inspiration for this book, the reason I started to research the true causes of obesity, was the astounding effect that bariatric surgery has on people's lives. I witnessed hundreds of patients, who had suffered with obesity for years, become transformed after the surgery. One of the most rewarding aspects of being a bariatric surgeon is the post-operative follow-up clinics, where we check on patients months and years after their surgery. It's a great feeling when a patient says, 'I owe my new life to you.' I often come out of the clinic with shiny packages of wine or chocolate to be redistributed to my long-suffering administrative staff (who spend stressful hours on the phone with patients who have had appointments and operations cancelled because of the inefficiencies of the NHS).

Quite often, if I have not seen a patient for six months after surgery, I will not recognize them when they walk into my clinic. The significant weight loss, combined with a change in their confidence and demeanour, mean that it's only when they show me their old photo that I remember who they are (or were). When I see them for their annual follow-up, years after surgery, it's usually just for a pleasant chat – they have changed their eating behaviour and enjoy cooking healthy, nutritious food.

We have learned that it is impossible to sustain weight loss in the long term unless you have first reset your weight set-point to a lower level. The basic premise of this book is that we can only do this by adjusting the type of food we eat, changing our food culture, de-stressing, improving our sleeping habits and maintaining good muscle health. But what is it about bariatric surgery that makes it so successful in drastically lowering the set-point?

We know that the hypothalamus, the part of the brain that controls our set-point, receives altered signals after bariatric surgery. The hormones that control our appetite and satiety drives are changed. These signals come from the gut. Bariatric surgery works by changing these signals by changing the configuration of the gut.

Gastric Bands, Gastric Balloons and Jaw-Wiring

In the early days of bariatric surgery, we thought that weight loss was achieved by one of two methods: either by a restriction of the amount of food that could be eaten, or by causing food to be malabsorbed. But we have now learned, through the usual medical mechanism – trial and error – that this is not the case. The gastric band (a plastic ring which is placed at the top of the stomach to stop you eating too fast), the intra-gastric balloon (a plastic balloon which is inflated in your stomach), or the old technique of jaw-wiring (where a dentist will literally clamp your teeth together) have all been shown to have poor long-term outcomes.

The weight set-point of obese patients has not been changed by any of these techniques; they merely create an obstacle to the food being consumed. If you declare war on the set-point by having one of these procedures, then you will undoubtedly win the first battle. Yes, you will lose some weight. But then the set-point will take control to stop you losing too much weight, your metabolism will collapse and you will crave the high-calorie foods that will pass through the mechanical obstacle to your guts that your doctor recommended. It is sad to see tearful patients who have had these procedures blaming themselves for poor self-control. Commonly they will regain weight by consuming soft ultra-high-calorie foods like chocolate shakes or ice cream. The formidable set-point defence has driven their appetite and food-seeking hormones through the roof – such signals (the same ones are produced after dietary weight loss) are too strong for us to ignore. The changed eating behaviour seen after weight loss in these patients, many of whom did not have a sweet tooth before their procedure, is driven by the change in these signals – not by any character flaw or weakness. This is the reason the gastric band, the intra-gastric balloon and jaw-wiring are increasingly being recognized as sub-standard – they do not alter the set-point and, as we now know, the set-point always wins the war.

Malabsorptive Procedures

What about malabsorptive procedures? We know that if you remove half of someone's bowel they will initially lose weight, but after a while they will automatically adapt to their shorter intestines by eating more. Their weight will eventually settle back at its set-point. The gastric bypass was initially thought to work by producing malabsorption, but we now know that this is transient – the smaller bowel adapts by becoming more efficient.

Sleeve Gastrectomy and the Gastric Bypass

There are currently two main types of bariatric surgery that really do work, and they work by permanently changing the weight set-point: the sleeve gastrectomy and the gastric bypass.

Both these procedures dramatically change the appetite and satiety hormones that we discussed in chapter 4. Ghrelin, the appetite accelerator, decreases considerably. This is the hormone that tells you to start looking for food when you have missed a meal – and the longer you go without food the stronger the signal gets. Eventually it will drive you to get any food with high calories – and it will make that food taste extra nice.

Peptide-YY (PYY) and *GLP-1** are the two hormones that control satiety, i.e. the appetite off-switch. These hormones are elevated to very high levels after both the sleeve and bypass procedures. The combination of high satiety and low appetite signals means that after this type of surgery a patient's behaviour will not be controlled by food any more – even if they have developed the leptin resistance that was described in chapter 5.

---

* GLP stands for 'glucagon-like peptide'. Just like PYY, it is released into the bloodstream by the small bowel after eating. It travels to the hypothalamus and causes the satiety feeling that is the signal to stop eating. As well as producing satiety, it has a second effect of making insulin much more efficient. This is the reason Type 2 diabetes often goes into complete remission immediately after a bypass (or sleeve) operation.

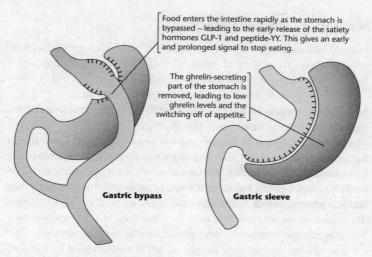

Food enters the intestine rapidly as the stomach is bypassed – leading to the early release of the satiety hormones GLP-1 and peptide-YY. This gives an early and prolonged signal to stop eating.

The ghrelin-secreting part of the stomach is removed, leading to low ghrelin levels and the switching off of appetite.

Gastric bypass

Gastric sleeve

Figure 6.1 The gastric bypass and the gastric sleeve

Bariatric surgery has become much safer over the last few years, as technology has advanced. I would now equate the risk of this type of surgery with having an operation to remove gallstones. Most patients will only stay in the hospital for one night and will be back to their day-to-day activities within a week of surgery.

If you are overweight or in the early stages of obesity I would *not* recommend this type of surgery. The guidance and suggestions in this book should be enough to help you reset your lifestyle, and sustained weight loss and a better quality of life should follow. However, if you are suffering with full-blown obesity and have developed leptin resistance or Type 2 diabetes, then this type of surgery may help you. Even if you follow all the weight set-point management strategies in this book, your body's leptin resistance may block you from achieving a significant reduction in your weight. For those people I find bariatric surgery is a life-changing procedure.

It is sad that we have come to a point in our human history where we need to develop more and more ways to treat man-made diseases. Bariatric surgery is one of these treatments. The surgeons who are trained to carry out these procedures are few and far between and yet the obesity problem is overwhelming. We are like a small group of firefighters rushing around putting out forest fires. But unless we help stop the cause of these fires, our efforts will be mostly futile.

My Typical Patient's Story

I would like to finish Part One of the book by focusing on what happens to a typical patient of mine. This story is an amalgamation of the hundreds of interviews that I have had with patients in my clinic over the last decade. Most of the stories are quite similar, so it is easy to summarize their struggles with obesity over the years, and then explain everything in terms of metabology.

My typical patient is female (80 per cent of patients undergoing bariatric surgery are female). She is in her forties and describes several members of her family who are also suffering with obesity (as we learned, 75 per cent of someone's size is predetermined by genetics). She has been obese or overweight since her schooldays and says that the school nurse was the first person to put her on a low-calorie diet. The diet worked transiently and she lost some weight; however, after a few weeks her metabolism caught up with, and adapted to, her low-calorie intake. Eventually, despite complying with the diet, she found that she was not losing any more weight as her metabolism matched her calorie intake. She felt tired and hungry and irritable and could not concentrate at school. After no more weight loss she decided to stop the diet as it was not working any more. It is at this point that she started putting the weight back on rapidly, as her low metabolism and voracious appetite helped her body regain its desired weight set-point.

She was worried that when she regained her weight it did not settle again at its previous level; on the contrary, she ended up with an even greater weight than before the diet. Her subconscious brain had calculated that she now lived in an environment where food was not predictable and therefore there could be another famine (or diet) around the corner. It is for this reason that her weight set-point now shifted upwards.

As the years went by, our typical patient tried all the different types of diet on offer (she mentions Slimming World, LighterLife, the South Beach diet, the red and green diet, the cabbage soup diet, Rosemary Conley . . . the list goes on). The diets are all different, but for our patient the result was usually the same: transient limited weight loss, followed by metabolic adaptation to the diet and a decision to stop it; then weight regain, and after each of the diets a new higher weight set-point.

Eventually our patient reaches a level of obesity where her fat cells cause an inflammatory reaction in her body. The inflammation

Figure 6.2 After dieting, a new weight set-point is established

stimulates insulin resistance, leading to an increased insulin level, and the higher insulin then causes the dreaded leptin resistance. A combination of the evolving leptin resistance and the legacy of previous dieting on her appetite (increased) and satiety (decreased) hormones means that the struggle with her weight gets more difficult the bigger she gets and the more she tries to diet.

This is a typical recurring story of initially successful diets, then weight regain, followed by yo-yo weight fluctuation through the years and, despite the constant conscious battle to diet, an inexorable rise in weight until serious end-stage obesity is reached. It is only at this point, after years of effort and sacrifice, after years of receiving the wrong advice from doctors and dieticians, after years of being misled by the food industry into the health benefits of bad foods, that my typical patient will tearfully admit to failure and blame herself for it. Finally, she will give up on her fight with obesity: many battles have been fought, but the war has been lost. The subconscious brain has won.

We have seen that you can't fight against your weight set-point by dieting – the only way to beat it is to understand it. We now know how the set-point works to keep your weight at a desired predetermined level, even if you are over-eating or under-eating. And we are now aware, from chapter 2, of the genetic and epigenetic factors involved in set-point calculation. But obesity, even in those people who are genetically primed for it, is not triggered until you are exposed to an obesogenic environment. In Part Two, we will learn how humans came to construct an environment so unsuited to them.

PART TWO

Lessons in Obesogenics
How Our Environment Determines Our Weight

SEVEN

The Master Chef

Why Cooking Matters

When I get home in the evenings I often find my teenage daughters watching TV programmes like *MasterChef*, *The Great British Bake Off* and *Ramsay's Kitchen Nightmares*. I don't understand their fascination with watching cooking and baking programmes, especially, I joke, when I never see food like this coming out of our kitchen. But I'm in the minority. Why do most people, just like my daughters, show an interest in these programmes? Why, in supermarkets, do you see crowds forming around the demonstration of a new gadget to make courgettes into spaghetti or to cut a cucumber into spiral shapes? Why are we reassured by watching someone chopping vegetables and cooking a meal in front of us – whether it be at home or in a new trendy teppanyaki restaurant? Why is the media full of recipes, and restaurant reviews, and articles on the latest new 'superfood'?

Why are humans just crazy about all things food? Once you know the answer to this question you will possess an important piece of the puzzle of obesity. I will demonstrate in this chapter why food selection, preparation and cooking define us as humans – and why, without fire and cooking, we could never have evolved into the intelligent beings we are today. This little-known secret explains where we came from and where we are heading in the future. It also explains the food-orientated, obesogenic world we have built around ourselves today.* It's all about getting enough energy to evolve, from the beginnings of life to the present day. Let me explain.

---

* 'Obesogenic' means 'causing obesity'.

The Replicants

In order to understand who we are now, at this moment, we must take a journey back in time to the origins of life on Earth. Think of a dark, stormy, tropical sea 4 billion years ago, before there was any oxygen in the atmosphere. Simple carbon-based chains of chemicals drifting around this primordial sea found, by chance, a formula to make new copies of themselves. These long chains would attract other chemicals that were floating in the sea until a double chain had been formed. The double chain then separated into two single chains which then continued the process of duplication. These were the first 'replicants', the chain of chemicals consisting of a primitive form of DNA. These ancient replicant chains turned out to be quite successful. They were able to coordinate the construction of more and more complex structures around themselves, eventually becoming one-celled organisms (think bacteria). Within the protective wall of these cells lived the replicant DNA code – the master controller – always coaxing and guiding the cell to further its spread. The survival of the code of life was all that mattered. Richard Dawkins, in his book *The Selfish Gene*, eloquently describes these organisms, constructed around themselves by the DNA, as survival machines: expendable biological vessels whose function was simple – grow, survive, reproduce.[1]

The Making of ATP Batteries

Our one-celled ancestors had a slight problem, though: they didn't have enough energy to grow bigger. They had developed fantastically efficient little *micro-batteries* – millions within each cell – that carried energy from food at the cell surface and dumped it into whichever part of the cell that needed to grow or move. These 'machines' (called ATP (adenosine triphosphate) in medical language) worked by charging up on food energy, then moving to discharge this energy within the cell, thus converting food energy into the type of energy currency that cells can understand and use. But the amount of energy the cells could produce was limited because they could not process oxygen, stunting their development into more complex organisms. Our one-celled ancestors were stuck in this evolutionary jam, remaining single celled for 2.5 to 3 billion years . . . Finally a solution came along – a development that would super-power our ancient cells and still, to this day, powers our metabolism.

New Tenants

A new type of bacterium was using oxygen to help it produce energy (oxygen started to appear in the atmosphere 3 billion years ago). This tiny bacterium had a unique internal corrugated membrane (basically a turbo engine) that enabled many more micro-batteries to charge at the same time. To our slow primitive one-celled ancestor these bacteria were like power stations – taking in and converting massive amounts of energy. How could they compete with them? Well, they didn't – they worked with them.

The new super-charging bacteria were either ingested (but not digested) by our one-celled relatives or the bacteria smuggled themselves inside our cells as parasites. Either way they survived and thrived inside our energy-deficient cell. It was a mutually beneficial relationship: our cell protected them and they produced lots of energy for us. They became endosymbionts, or a cell living inside the cell of another organism.

These primordial super-charging bacteria did well out of this alliance. They are still part of us and of all animal cells today. Our bodies are powered by them. They have become a vital part of us, helping to convert complicated food energy into cellular energy (or heat). These cellular power stations are called mitochondria, originating from those primordial bacteria and assimilated within us.*

With our energy-generating capacity super-powered by our helpful bacterial tenants, more and more complex organisms developed and evolved until the present day: we now have an estimated 10 million species on Earth. But for all the diversity between these life forms – fungi, plants, fish and animals – they all have one thing in common: their DNA originates from that single replicant template in the primordial sea. Ninety-nine per cent of all the species that ever lived on Earth became extinct, but those of us that are left are the modern dynamic survival machines – controlled by our DNA bosses, powered by mitochondrial endosymbionts and directed to survive, grow and reproduce.

Since that first chemical replication of a simple protein in the primordial sea, there has never been a dropped generation. In those 4 billion years there is a 100 per cent success rate in terms of growing

---

* Mitochondria are like little furnaces within our cells, constantly generating heat and power. Our metabolism, that is the amount of energy that we can use, is dependent on these furnaces.

and surviving long enough for each generation to reproduce and pass the master DNA code on to their offspring. Generation after generation evolved and adapted to the changing landscapes and environments of the Earth. With our complex family tree we have accumulated 4 billion years of ancestral heritage, or baggage, in our genes, and this has moulded who we are now and how we survive.

In the same way that an artist adds successive layers of paint to the canvas to finish their masterpiece, so we, as humans, contain deep layers of evolutionary history that cannot be changed: this work of art took 4 billion years to complete, with each evolutionary change adding a fresh layer, a new look.

The Energy Budget

Every living organism today is related to our common one-celled ancient ancestor, and that means every organism uses the same system of energy to survive and thrive. Bacteria, plants, algae, fungus and all animals from snakes to birds to humans – we all have those *ATP batteries* converting food energy into usable energy for cells. Even viruses use ATP batteries (but not their own – they borrow the ATP from whichever cell they have invaded).

Our primordial energy rules provide every animal with a maximum amount of energy that it can use per day. This is called the energy budget. The bigger the animal, the larger the energy budget. But unfortunately, a budget is exactly what it means – a limit to resources. Evolution has to budget enough energy to keep all the organs happy in each species: enough of a balance to keep the animal (or survival machine) alive, the heart pumping, the lungs breathing, the muscles working, the stomach digesting. But one organ needs more power to run than the others. This shining beacon of energy extravagance is the organ that differentiates humanity from all other species – the brain. How could we, confined to a limited energy budget, afford to make our developmental leap with a big, energy-hungry brain? The answer to this evolutionary puzzle explains why we humans love certain types of food.

Chimps Don't Get Fat

Fifteen million years ago our close cousins the chimpanzees developed from gibbons. Chimps are obviously still around today, so we know that they stay in the rainforest mostly and forage for fruits, nuts, insects and occasionally meat. As an aside, in many rainforests where chimpanzees live there is a total abundance of foods all year round. Chimps can gorge on this type of food for as long as they like, and yet, even with plentiful food, wild chimp populations never develop weight problems.

About 1.9 million years ago some chimpanzees started behaving differently. They started walking on their hind legs for longer and longer periods of time. Because of their new upright stance their distance vision improved and they were eventually able to leave the rainforests on two feet and roam the savannahs, hunting and exploring and inhabiting new areas of the world. As time passed they grew in height and developed into extremely efficient runners with more stamina and endurance than other animals – until they were able to run exhausted prey down. This meant more success in hunting and therefore more meat and protein. This species was called *Homo erectus*.

Then came the biggest shift of all, the shift from the smaller-brained *Homo erectus* to the large-brained humans that we are. This occurred about 150,000 years ago when the first anatomically modern *Homo sapiens* evolved. The process of evolving a large brain in a new species is called cephalization (from the Greek word *enkefalos* for brain). It might be worth reminding ourselves that our large brain made us not only clever, but vicious – so when we did eventually branch off from our brother *Homo erectus*, we killed them, all of them, along with our slightly dimmer but stronger Neanderthal cousins (whom we still share a part of our DNA with).

The Expensive-Tissue Hypothesis

How could humans have afforded to develop a brain four times larger than our ancestors – an organ that used up so much energy? We could not break the energy rules embedded within us; we would have to start from scratch and go back 4 billion years. One of our other organs would have to be sacrificed in order to free up this energy within our confined budget.

Evolutionary scientists had been debating this question for years with no agreement on how it could be done until a research paper entitled 'The Expensive-Tissue Hypothesis' by the anthropologists Dr P. Wheeler and Dr L. C. Aiello gave us an explanation.[2] The paper began by calculating the energy budget of animals according to their size. The amount of energy that an animal can use up over time is called its metabolic rate. This is the same as the amount of power that it needs to function.*

Think about power, and the amount of power different animals require in order to function. Imagine that food energy did not exist and animals needed to be plugged in, like electrical appliances, for them to work. In mammals the amount of energy or power needed to live is dependent on their weight. A dog will need much less power to function than a 65kg human – unless the dog happens to be a 65kg Irish Wolfhound, in which case it will have the same energy requirements as the human! It is all about the total number of mitochondria that each animal has in its cells – these are our biological engines and determine how much cellular energy our micro-batteries can produce per second.

The Chancellor's Cuts?

Imagine that the Chancellor of the Exchequer, in his annual budget, had to look after a human body rather than the finances of the British economy. As he held up that battered old briefcase to the press outside Downing Street, how would he go about the strategic planning to turn *Homo erectus* into *Homo sapiens*? Instead of budgeting for the different departments – Health, Defence, Environment, Transport, Education – he had to budget for different vital organs: energy for the Heart, Lungs, Intestines, Muscles and the Brain. Instead of making unpopular cuts, for instance in Health and Defence, in order to make the country brighter in the future by expanding Education, how would he find the resources to expand brain size fourfold? Which organ could he afford to trim down to find the spare energy in his constrained budget to run the bigger department of the Brain? Could he afford a cut in Muscle, or Heart or Lungs? Surely this would affect our chances of surviving predators and finding food.

When some anthropologists looked at the size of human organs

---

* Power is measured in watts, which we usually relate to electrical appliances, but it applies to all living and moving things. It is the amount of energy used per second. The power might come from electricity in the case of a washing machine, or a petrol engine in the case of a car, or food (then ATP) in the case of animals. To get things into perspective, 1 joule of energy is the amount used to lift an apple out of a 1-metre-deep barrel; 1 watt is the power required to lift that apple 1 metre in 1 second. To lift ten apples (or about 1kg) up 1 metre in 1 second takes 10 watts of power.

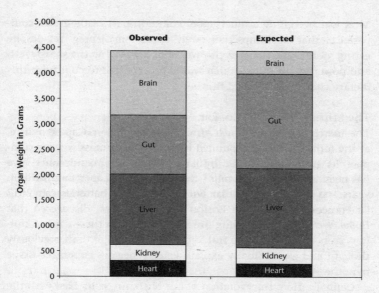

Figure 7.1 The density of brain mass in humans is much higher than would be predicted for a primate of our size and the GI tract is much smaller than predicted

Source: L. C. Aiello and P. Wheeler (1995). The expensive-tissue hypothesis: the brain and the digestive system in human and primate evolution. *Current Anthropology*, 36(2), 199–221.

and compared them to the size of other similar primates, they noted that humans have a much larger brain and a much smaller gut than similar-sized primates. Their conclusion therefore was that the metabolic room necessary for brain growth and evolution came from a decrease in the size of our guts – to evolve our large brain, our gut was sacrificed.

But how could such a radical cut in the size of our digestive system occur without it affecting our health, or our wellbeing as a species? Would this not cause us to starve or become malnourished? The answer lay in the environment and not within our bodies. Our closest ancestors, *Homo erectus*, had already begun to develop a brain that was larger than the chimpanzee's. Archaeological digs have uncovered evidence that they were using knives made from razor-sharp flint to cut up meat. We also know that their natural stamina and increasing cunning made hunting expeditions more successful.

Meat was becoming a much more common part of their diet compared to that of chimps that rarely caught small prey. Yet despite eating more meat, *Homo erectus* were not developing the sharp teeth and powerful jaw muscles that would be expected to cope with this dietary change. How could this be?

The Million-Year-Old Cooker

The answer came from South Africa. In the Northern Cape Province, at the foot of a hill, surrounded by scrub, lies a massive cave complex, its entrance obscured by large stones. The Wonderwerk cave has been inhabited by humans, pre-humans and apes for 2 million years. It is one of the oldest known sites of human habitation. In 2012 Dr Francesco Berna from Boston University, USA, discovered that *Homo erectus* had been using fire to cook food in these caves 1 million years ago.[3] This was a full 200,000 years earlier than previously thought and would finally explain Peter Wheeler's expensive-tissue hypothesis.

Cephalization, the evolution of our big brain, coincided with the discovery of fire (800,000 years prior to the first humans, i.e. plenty of time for our species to develop from *Homo erectus*) and with our increasing mobility and improving vision. The discovery of cooking meant that a much wider variety of foods could be eaten.

If *Homo erectus* could make and control fire 1 million years ago, then this would explain why, despite the increased consumption of tough meat, their teeth and jaws actually became smaller. The combination of the control and use of fire and the availability of different types of foods, plus much more meat, led the developing humans to start cooking. They cooked their meat to make it easier to chew and swallow. In addition, vegetables were starting to be cooked, meaning that easier digestion of roots and tubers (like sweet potato or cassava) was now possible.

Our ancestors cleverly used the energy in fire to cook foods, to break them down and make them easier to digest. Raw foods take more energy for our digestive systems to process than cooked foods. This is because the very process of cooking is almost like a pre-digestion. Cooked foods required a less effective intestinal tract to extract the same energy compared to raw foods. The discovery of cooking was the most important factor in conferring a distinct evolutionary advantage for humans over other species. Cooking meant that the quality of our food improved and that we did not need such a long gut to digest it. As the size of our gut decreased, we were left with metabolic capacity in our energy budget to develop a larger brain. We are human only because of this development.

A Chef, a Chimp and a Gorilla

As an example, imagine going to the zoo and in one of the enclosures seeing a 65kg man. Imagine a slightly smaller version of chef Gordon Ramsay, or whoever your favourite TV chef is, to make it more vivid – he is standing at a stove frying steak and eggs – and swearing a lot. Next to him stands a 65kg full-grown adult male chimp (eating a mound of nuts and fruits), and on the other side of him is a 65kg growing young male gorilla (eating bamboo and termites). All three of them have the same energy budget per day because they are all mammals and weigh the same. They all need about 2,000kcal per day.

What is the difference between the three? If you weighed the sizes of their hearts, livers and kidneys, you would find they were similar. However, the reason the chef is able to articulate many swearwords while carrying out complex cooking tasks is because his brain is four times larger than those of the guys next to him making the odd shriek and grunt. The chimp and the gorilla have a much bigger gastrointestinal (GI) tract than the chef – because they eat raw food all day. The chef on the other hand has evolved to have a smaller GI tract because much of the energy that the chimp and gorilla use digesting raw foods he has saved by cooking his food. Not only that but he can take in all his daily nutritional requirements much faster than the chimp, and particularly the gorilla, who spend large parts of the day eating. The release of energy from cooked foods not only enhanced our ability to evolve larger brains but also gave us the time to use that brain while other animals were eating.

This is an example that I hope you remember because I wanted to demonstrate that when you look at the differences between those three 65kg primates (yes, the chef is classified as a primate – as are you), the difference is the cooker in front of him. Without roasting, frying, boiling or baking our food we would not have been able to evolve a small gut and a large brain. Cooking food is a very big part of us, and is part of what has made us human. This is why we are still fascinated by all things connected to food and cooking – it is as fundamental a part of us as those foreign bacterial furnaces, those mitochondria, squatting in our cells and keeping us warm and alive.

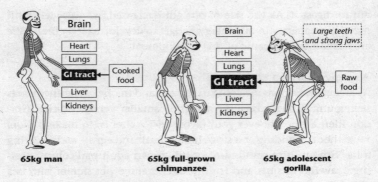

Figure 7.2 Comparison of size of brain and gut in 65kg man with equivalent weight chimpanzee and adolescent gorilla

Source: Adapted from the drawings of Mr Waterhouse Hawkins from specimens in the Museum of the Royal College of Surgeons.

Could We Survive on Raw Food?

Now that we have developed a smaller gut, is it possible for us to go back to the days before fire and survive on raw foods only? Are we now reliant on consuming foods that have been cooked and 'pre-digested'? Can we go back to raw food, or is cooking now a fundamental part of us?

Devotees of raw food seem to think it is possible to return to pre-fire days. Raw foodists think that by eating raw foods only they will have increased energy and health. Let's look at the evidence of a study of over 500 raw foodists living in Germany in 1999.[4] The study found that when they switched from cooked to raw food they lost a considerable amount of weight. Women lost an average of 12kg and the men lost an average of 10kg. A third of the group had evidence of severely low body weight with chronic energy deficiency. Of the females, 50 per cent reported that they had stopped menstruating and had therefore become temporarily infertile (their low leptin levels caused a shut-down of fertility as a safety mechanism). These quite worrying health outcomes occurred despite the raw foodists having all the benefits of living in the twentieth century. The raw foods that they could select from the supermarket or grocer were abundant and wide-ranging and, unlike in the wild, there was no seasonal deficiency of certain foods. The raw foodists of today could even select prime cuts of raw salmon and steak tartare. They consumed high-quality olive oils which gave them 30 per cent of their total energy intake. They were able to use blenders, to finely slice, grind and liquidize their ingredients and make them more digestible. They had all the advantages that modern society gives us. But despite this a third were found to be severely malnourished and half of the women were infertile.

Without the advantages of the modern world a hunter-gatherer tribe without fire and cooking would fare very badly in comparison. If the raw foodists had been a tribe or community in hunter-gatherer times they would have dwindled to extinction within a few short generations.

150,000 Years BC

Let's take stock of our journey from the primordial sea to *The Great British Bake Off* show. The time is 150,000 years BC. We have a shorter gut compared to chimps and other monkeys and are reliant on cooking foods for our health and the survival of our species. But as the gut shrank, so our brain size increased, and gradually we evolved fully into what is described by anthropologists as an 'anatomically modern' human. What does 'anatomically modern' mean?

Cro-Magnon Man in Regent's Park

If you took a *Homo sapiens* (Cro-Magnon man) cave-dweller from around that time and transported him to the twenty-first century, gave him a wash and dressed him in jeans and a shirt and sat him on a park bench – no one would look at him twice. He would have a dark complexion and blue eyes.[5] He might be weather-beaten and have calloused hands and he could be eyeing up the squirrels suspiciously. He would have an encyclopaedic knowledge of nature, the seasons and the stars. He would be a dedicated family man, nurturing and protecting them with his life. He would probably be taller than the average man now. But otherwise, from the outside, he would be just like us. No one would notice any difference. We might think he is a tourist from southern Europe.

What would amaze us though, if we could look inside this man, would be a complete absence of any type of modern disease. His

Figure 7.3 Forensic facial reconstruction of a Cro-Magnon man, using a cast of the skull

Source: This file is licensed under the Creative Commons Attribution-Share Alike 4.0 International License. Used courtesy of Cicero Moraes.

heart would be pristine, with no signs of atheroma damage,* and his blood pressure would be as low as an athlete's today. There would be no sign of any type of inflammatory condition such as arthritis and no asthma. He would not be diabetic, and most striking of all there would be very little chance that he would be obese. His body weight would almost certainly be well within the normal healthy range. If he could access our healthcare system for treatment of the conditions that he would have been susceptible to (mainly trauma and infection) he would probably live beyond his ninetieth year.

What did our bushmen ancestors eat to keep so healthy? We know the 'Paleo diet' (short for 'Palaeolithic', the epoch the cavemen lived in) is supposed to mimic the diet consumed by our ancestors, but what did the real hunter-gatherers live on? We can find this out by studying current hunter-gatherer populations which still remain isolated from modern influences: the Hadza tribe of Tanzania, the Bushmen in Namibia, the Pygmies of the Congo jungle, isolated Amazonian tribes, the Inuit Eskimos in Greenland and the Aborigines in Australia.

---

* Atheroma is the 'furring and narrowing' of blood vessels that affects most adults living in the West. It can lead to heart attacks, strokes, kidney failure and many other Western diseases.

The Hunter-gatherer Supermarket

Let's go to the hunter-gatherer supermarket, a vast open-roofed store where all the produce is free, there are no cash tills, but you have to pay in time and energy to fill your bag. There are two sections to this store: one is the greengrocer's, for fruits, vegetables, nuts, fungi (like mushrooms), eggs, snails (or shellfish), green leaves and herbs; the other one is the butcher's: for meat. In both sections, there will be small honey areas. Only women and children shop at the greengrocer's section and only men go to the butcher's section.

The supermarket's greengrocer's section is quite large – 2 square miles (5.18 square kilometres). The produce is spread out and hidden in bushes or under rocks or soil, meaning that it can take many hours for a woman to find enough food for the evening meal. The most popular carbohydrate of the diet, the food that they rely on the most, is the roots of a plant, the part where all the dormant energy is hidden below ground to keep it safe from leaf-eating, foraging animals. The buried food treasure survives whatever the season; it is available all year round, and is therefore a reliable staple food. The women will have brought special sticks with them to uproot these tubers, roots and bulbs (think sweet potatoes, yam, cassava, ginger and certain flower bulbs – which, when cooked, are safe and nutritious). They will also look for seasonal foods that are found above ground, such as berries, fruits, seeds, mushrooms, nuts, edible flowers and leafy greens and shoots. They will gather bird eggs, snails and, the biggest treat of all, honey.

Men Only

Now to the meat section of the hunter-gatherer supermarket: men only allowed. There is twenty-four-hour opening to allow some groups of men to camp out in the middle of the store overnight if they do not find what they want. The size of this section could be as much as 40 square miles (103.6 square kilometres)! It's interesting to see the techniques for obtaining foods. Groups of five to twelve young and fit men will see an animal from afar. Human males are not as fast as their prey, but they have two advantages. The first is that after learning to stand and balance on two feet, the early humans rapidly became the most efficient animals at moving. They lost their insulating hair as they discovered warming fires and clothes, meaning that they could cool themselves very efficiently by sweating, compared to most animals that rely on panting to cool their bodies when running. Humans take less energy per unit weight than any other mammal to move distances.⁶ The second advantage that our hunter groups had was their stunning brainpower. They were able to work in teams – communicating, planning and learning how to track down and trap animals. In the early days, before more sophisticated weapons like arrows and spears were used, the hunters would simply go on a marathon trek, keeping the prey in view, and eventually run it down with superior stamina before bludgeoning it with heavy rocks (a technique known as persistence hunting).

Because the meat section is open 24/7, it also contains a scattering of snack foods to keep the boys going for longer. Tasty insects, fruits and eggs are sometimes available – as well as the honey tree, which requires smoking to unlock the honey from the bees safely.

Back to the Camp Fire

Every day in the afternoon the women would return home to their camp with the food they had gathered. Usually the men would return in the evening. It might seem sexist to us that the hunter-gatherer supermarket had separate sections for men and women. This doesn't happen in any other species. Why are humans so different? Why did early humans organize their male and female tribe members to go looking for different foods? The answer again goes back to cooking. The main staples of their diets were root- or tuber-type carbohydrates (sweet potato, cassava) and wild meat. Both types of food needed to be cooked over a fire. The cooking would be done in the evening, although the cooking fire would be kept alive throughout the day and night. For all other animals, once the food has been killed or foraged, it is eaten raw by the animal that got to the food first (unless it is a mother weaning her young). With humans, the food would not be palatable raw and so had to be taken back to the camp and eaten later. This meant that within family groups food could be shared out among males, females and children.

In no other animal group is food commonly shared between males and females. The whole concept of cooking together and sharing meant that the members of the tribe who were more likely to be successful hunters could be deployed for this job (young fit men) and those members who would be less successful hunting (women, even if carrying a baby or young child) could use their energy to gather plant-based foods. In the evening the families would come together to share food and eat. This was essential for the survival of the tribe. Over the camp fire ideas and stories would be passed on from one generation to the next, so that knowledge acquired could be used by future generations. In this respect, it was not just the chemical energy of the fire to break down food that helped them biologically; it was the social structure around the fire, plus the cooking and the sharing of food among the families, that helped the early humans to continue learning and progressing.

The hunting party could be away for long periods of time. Once they had killed an animal, quite often they would eat the most prized part of it raw immediately, before carrying the rest back to camp. The most nutritious part of their kill was the liver. One of the characteristics of hunter-gatherer populations (something we have lost) was the consuming of the innards of an animal before moving on to the carcass. Hunter-gatherers would value the liver, kidneys, intestines, bone marrow and brains of an animal above its lean meat, as these organs contain much more nutritional and energy value than muscle tissue.

The Tasty Offal

The food quality of these organs, many of which we now commonly discard, is exceptional – they contain many essential fats, vitamins and minerals. Our hunter-gatherer ancestors prized one nutrient above all others – even above the sugar in honey – fat. Any food that contained high proportions of fat within it would be favoured. They would laugh at current expert advice (we will discuss this in more detail later in the book) that fat makes you fat. Our ancestors instinctively knew that fat was essential for making them strong and healthy.

If you look at the fat content of different animal organs compared to animal meat you can see why they would choose any type of offal first. Lean meat has a fat content of only about 5 per cent, compared to kidney, which has 15 per cent fat, the intestines (tripe) 18 per cent, the heart 25 per cent and the liver 30 per cent.[7] And the organ with the most fat? The brain – up to 50 per cent of brain tissue is fat; it contains high levels of the essential (and currently very misunderstood) type of fat we now call cholesterol. In addition to the animal's organs, the fat under its skin and inside its abdomen would be prized. One of the most highly nutritious parts of any animal is the bone marrow, the congealed tissue within the long bones of mammals that is responsible for making blood cells. Bone marrow is made up of 84 per cent fat (in all caveman excavations, there is evidence that animal bones and skulls would have been smashed to access these nutrients). There is good evidence that when hunters fell on hard times they would scavenge the bones of animals already killed by other predators to access this precious source of energy.

Vegetables, Fruits and Carbohydrates

You can imagine the difference in the quality of foods foraged and hunted by our ancestors compared to today's equivalent food types. The vegetables, fruits and tubers would all have grown wild and would not have gone through our terribly wasteful supermarket quality control. In today's world, up to a third of all fresh foods grown for supermarkets do not reach us because they do not measure up to the supermarkets' standards, because they either do not look right, or are bruised, or are just not fresh enough. This has not always been the case. Early man was used to a rich variety of wild fruits, berries, green shoots and root vegetables. They may not have tasted so sweet and ripe as today's hybridized, genetically engineered, odourless and perfect-looking equivalents, but the variety would have far exceeded what we consume today. In temperate regions over 100 different types of plant foods were consumed, and in tropical climates many more. The dietary amount of carbohydrate consumed by our ancestors was much lower than today and the carbs that they did manage to obtain were totally unrefined – certainly they would not have been the major source of food satisfaction that carbohydrates are for us today. Their equivalent might have been grilling to a crisp a piece of wild boar liver!

The Palaeolithic Diet

We will leave our hunter-gatherer friends for now and look at what happened to them as they evolved. But, before we go, let's consider a summary of the food they actually consumed. This shows what the real Palaeolithic diet consisted of: lots and lots of meat, fatty offal and bone marrow, with a top-up of a staple unrefined carbohydrate and seasonal foods as a treat.

Extending the Food Supply

So, what happened after the cave dwellers evolved into early humans 150,000 years ago? They spent a lot of time doing what they did best – roaming and colonizing the Earth. They became more proficient hunters and developed language and kinship.

When foods were plentiful and the weather was OK and they were in a safe place, we might assume it was like the Garden of Eden, a paradise on Earth. But the reality was very different. They had evolved

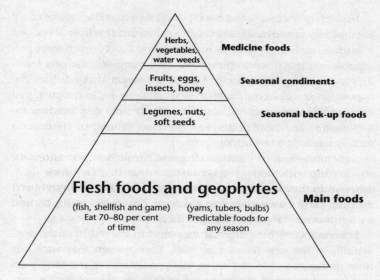

Figure 7.4 The hunter-gatherer food pyramid – if it had existed

Adapted from M. Sisson (2012). *The Primal Blueprint.* London: Ebury Press.

as humans because of their relationship and fascination with food. Food was obviously critical for survival, but cooking and preparing foods had also been instrumental in their ability to evolve and become 'human'. The problem was that hunted, or foraged, foods were unreliable. Foods were seasonal. Our ancestors had to trek to different areas depending on the migration patterns of the animals they hunted and the weather. They were constantly upping camp to find seasonal foods.

They started to solve their food supply problem in around 20,000 years BC. In a part of the world that is now Egypt, the first farmer to ever live discovered that some grass seeds (particularly early forms of wheat like buckwheat and spelt) could be planted and grown in the moist and fertile soil that is unique to the area. This significant development meant that their food supply could be predicted and controlled and the hunter-gatherer clans did not have to keep moving to find food: it was the advent of the agricultural age. As well as controlling plant food, they learned how to tame and domesticate animals so that herds of cattle, goats or sheep could guarantee an easy, all-year-round supply of meat.

Now that they finally had a stable food supply, our ancestors could put down roots and build permanent settlements that later developed into towns and cities. Farming made the food supply much more predictable and much more efficient. A small number of farmers could supply food for many people. Agriculture meant that much of the population of a town did not have to spend their day in the pursuit of food, unlike their nomadic ancestors. Their time was freed up for tool-making and later to develop the other benefits of civilization such as science and education.

Agriculture and civilization all sound great. Now our ancestors' relationship with food – the very relationship that had made them different to the chimps – was developing. After discovering fire and learning to cook food, they now had control of their food supply, and with this came the power to utilize their brains.

However, something unusual happened to the health of the populations in the new towns and cities. Even though they were safe from predators and from famine, the populations of the early post-agricultural age were weaker and shorter than their hunter-gatherer ancestors. Many became poorly nourished as they were now eating the limited foods that were available from farming, instead of the wide variety of plants and animals that the hunter-gatherers had consumed.

The impressive and inquisitive brains of the humans didn't stop evolving when they were able to harvest their own crops and butcher their own cattle. The agricultural age continued developing at an increasing pace as communication and transport between towns improved. Clay pots were created to cook with, as well as to eat on, and store food in. The wheel used by the potters was spotted by someone with an inquisitive mind and adapted to make the wheels of the first carriages, kick-starting a transport revolution. The wheel principle was also used much later to build mills, often powered by river water, to help grind wheat down. Iron was used to make useful farm tools like the plough. Irrigation, dams and crop rotation were developed.

As the productivity of farms improved, so farmers found that they had an excess of food. More food than they needed to feed their family and neighbours. They started to trade and sell their wares in local markets, increasing the types of food available to them.

With improved transport, local markets which would serve an area of a 10-mile radius soon progressed into national markets. Merchants became involved in trading between local markets, purchasing foods in bulk in one market and selling it at a profit in a distant area where there was a need. This development is important because it heralded a shift to people no longer eating products that were grown in their own local area. The profits that merchants made from trading food and transport were good. The people were also happy that they had a variety of foods from many miles away to choose from. But the genie was now out of the bottle – and the results of this new trading economy would seal our future relationship with food, with unexpected consequences.

THE VICTORIAN GIRL COMING OF AGE

Imagine London in Victorian times, maybe around 1850. A mass of humanity had descended on the city from surrounding towns and villages – poor people with a dream to make their fortune, but in reality just about surviving. Dirt, noise, disease and crime fill the rat-infested streets. But there is another side to this Victorian scene.

In exclusive Arlington Street, overlooking a lush Green Park, a young aristocratic lady sits at her dresser and makes the final preparations for her coming of age celebration. She smiles at herself in the mirror, admiring her wealth and status. She glows with excitement. The mixture of powdered iron filings, water and vinegar that she had rubbed on her teeth had done the trick – they had turned black! The culprit of this seemingly inexplicable fashion faux pas? Sugar.

The Irresistible Pull of Sugar

Black teeth as a fashion statement? This was a sign of our evolutionary Achilles heel coming to haunt us. There is a lot of skipping backwards and forwards through time here, but I'm afraid we must go all the way back to the ancient savannah now to explain our weakness for sugar. Evolving humans, with lots of new environments to explore, had to develop a safety mechanism when choosing new foods to eat. They needed to know what was nutritious and what was poisonous – and to facilitate that our ancestors developed sensors in the mouth that would give them clues as to whether something was safe to eat or poisonous, and whether it was nutritious. These sensors are still with us today: they are the taste buds located on our tongue. There are six types of taste that we can differentiate: bitter, sour, salty, fat, protein (called umami) and sweet. Any food with bitter or sour flavours would make us cautious before eating it. Foods with mainly salty, fatty or protein flavours would get the 'OK' signal. But any food that lit up hunter-gatherers' sweet taste buds would also light them up too.

Evolution has hard-wired our sweet taste buds straight to the pleasure area of our brains. If the signal is strong enough – if the sweet food is sweet enough – the signal we get, straight to our brain, is the same as if we had taken an opiate drug like morphine or heroin (maybe not a large dose – but the signal is the same). The sweetness signal calms our emotions and improves our mood.

In fruit, the sweet taste is a signal from the plant to the animal (or us) to eat it. If the fruit gets eaten, its seeds get propagated far and wide. Sweet-flavoured foods obviously contain glucose and, as we know, our brains are expensive to run. They need a constant supply of glucose to work – otherwise we will fall into a coma rapidly. This explains the importance in our evolution of prioritizing sweet foods by making us go crazy for them.

In hunter-gatherer times, sweet foods were very hard to come by. They tended to be seasonal like fruits and so our ancestors' taste for sweet foods prompted them to trek far and wide to find them. But at this time hunter-gatherers only had access to the great feeling during the summer seasons. That was until farming and transport and our friendly merchants stepped in.

In most parts of the world farmers grew a staple food for their population, depending on the climate. In North Africa, the Middle East and Europe, that staple was wheat. In India and China, it was rice, and in America it was originally maize. Yet none of these staple foods gave the early humans the 'high' they craved when they consumed sweet food.

Then, 10,000 years ago in Indonesia, farmers first cultivated a type of stout grass that accumulated sugar in its stalk: sugar cane. People loved to chew and suck on the cane to release the sweet juice. Sugar cane farms soon spread across Asia. But sugar cane, unlike wheat or rice which contained the nutritious part in the seeds and could therefore be stored, needed to be consumed soon after picking, otherwise it would rot. This meant that trading sugar cane on anything other than a local scale was not possible. Then a breakthrough came in India around AD 300. Farmers discovered that if the sugar cane pulp was squeezed or crushed out and left to dry in the sun it would form solid sugar crystals, meaning that sugar could be processed into a commodity fit for transport and trade. It became a valuable 'spice' used in cooking and in medicine to treat illness (or, in actual fact, to make ill people feel better with the opiate effects of sugar).

In the Middle East, techniques to refine sugar progressed and it became an integral part of Arabic culture. Delicious sweets were produced that were revered by anyone tasting them. European civilization was not exposed to sugar until much later. Probably the first contact with sugar traders was during the Crusades of the eleventh and twelfth centuries. French, Roman and English soldiers started to bring 'sweet salt' back to Europe, stimulating the interest of royalty and other wealthy citizens. Spain, Cyprus and Portugal (Madeira) started to produce their own sugar, but the price remained extremely high due to intensive labour costs in growing and processing it. It remained a rare, expensive delicacy.

Sugar and Slavery

Then the story of sugar gets darker. On discovering the Caribbean islands in the late fifteenth century, early explorers noted that the climate would be ideal to grow sugar cane. Within a few years, the first Caribbean sugar plantation was established in Cuba in 1501. The demand for sugar in Europe was extraordinary and merchants saw an opportunity to make huge fortunes – but they desperately needed people to work on the labour-intensive plantations and sugar mills. They turned to the slave trade and eventually 10 million Africans were forcibly transported to work on the sugar plantations of the Caribbean and Brazil. The merchants lined their pockets with every transaction – filling slave ships in West Africa and selling the slaves to plantation owners in the Caribbean, then loading the ships with sugar (and rum) in the Caribbean and selling it in Europe. Finally, they would complete the triangle by transporting guns and munitions from Europe and trading them with African warlords for the slaves they had captured from neighbouring tribes. It became a highly profitable trade in misery.

The Sugar Glut

By the 1700s and 1800s the glut of sugar being produced in the Caribbean finally meant that sugar products were much more available to those living in the West. Sugar was seen as highly profitable by the government, which taxed the import of 'white gold' prohibitively. A pound of sugar cost 2 shillings or the equivalent of £50 in today's money.* It was now becoming a delicacy for the common working

---

* Until the twentieth century sugar was transported and sold in the form of a 'sugar

man, but a staple for the aristocracy. The excessive consumption of sugar by the Victorian aristocracy led to an excess of tooth decay in this group. The presence of black and rotting teeth in a person signified that they had the financial means to have bought enough sugar to rot their teeth – at the time this was deemed a desirable look! Of course, if you were too young to have rotting teeth, the fashion was to mimic this affluent look by painting your teeth black.

> *As she peers out of the window of Arlington Street into the park, our young lady feels pity for the poor people down there – those who cannot afford to sample the delights of sugar, and certainly cannot afford to dye their teeth a fashionable black. But she should not feel too sorry for them. One of the hidden secrets of the Victorian poor was that they were, by chance and not by design, living in a nutritional golden age.*

The Victorian Miracle Diet

Average life expectancy in Victorian England was forty-one years. However, infant mortality for the poor was shockingly high and skews the average life-expectancy statistics. In poor working-class areas of England, infant mortality reached almost 25 per cent; in the slums it could be as high as 50 per cent. Most children died through infectious diseases, such as dysentery, cholera or typhoid fever, due to poor sanitation.

However, if infant mortality is excluded from the health statistics of the time, then the life expectancy of a poor Victorian, as long as they had made it to their fifth birthday, was similar to that of today.[8] Even without the benefits of modern medicine Victorian life expectancy mirrored ours.

The health of these poor Victorian populations (who had survived their early childhood) was down to the unique diet of the time. There was no food shortage. Fresh foods could be purchased from the markets relatively cheaply. The diet consisted of vegetables and roots, including onions, leeks, carrots, beetroot, turnips, Jerusalem artichokes and large bunches of watercress. In the summer cherries and

---

loaf' – a large conical-shaped lump of congealed sugar. The sugar loaf was prized and made to last for a long time: special pliers were needed to break pieces of sugar off the solid loaves.

plums would become readily available, and by autumn gooseberries and apples were in abundance. Dried fruit was a common treat for children. Legumes such as beans and peas were plentiful as was the winter delicacy of delicious roasted chestnuts.

Being an island nation there was an abundance of fish, including salted or pickled herrings, eels and shellfish such as mussels. Meat was less commonly consumed, but when it was, it was eaten in its entirety; just like our cavemen ancestors, the Victorians knew the health benefits of bone broth and certainly enjoyed the delicacies of offal: the heart, kidneys, 'pluck' (intestines and lungs) and, most importantly, the brain. Most of the meat they consumed was in the form of this cheap offal – rich in essential micronutrients and in saturated fats, especially cholesterol.

The Victorian diet of the poor was low in sugar and refined carbohydrates and high in fresh vegetables, fish and the health-giving properties of bones and offal. People smoked less frequently than today and beer was watered down, meaning alcohol consumption was also lower than now (the role of alcohol will be explored later in this book). Combine all this with an active job and soon you get to live the same age as we do now without having access to modern healthcare to achieve it. However, the healthy mid-Victorian diet would not last longer than a generation. By 1870 beet sugar from Europe was flooding the market and undercutting Caribbean sugar cane imports. The price of sugar had started to plummet, meaning that the golden diet of these Victorians would never return.

Pandora's Food Box

Just as, by chance, we had stumbled on the perfect, healthy Victorian diet, another massive change to our food environment occurred. In the Industrial Revolution farms became mechanized and therefore more profitable, transport became more efficient, and food became a business. For the first time in human history the population had access not only to locally produced fresh foods but also to foods from long distances away, sometimes from different countries or even different continents. This food had to remain edible, despite the long distances travelled. Ideally, food would be produced that could have a long-term shelf life. This meant that the food had to be altered to extract the parts that would make it go off (in most food this includes omega-3 'good fats' – we'll discuss this later) which would then be replaced with a substitute that could act as a preservative (various of the E-numbers you now see on food packaging) and other ingredients that would make the food more palatable (mostly sugar, salt and fat combinations).

Hybridization and Genetic Engineering

A critical factor in industrialization was the availability and the cost of staple foods. For example, wheat had come a long way since the advent of agriculture in Egypt. The natural wheat of those times had slowly been changed, first by hybridization techniques (where the two different strains of wheat could be combined to grow a stronger or taller crop), then more recently by genetic engineering. If you are now forty years old or more, and you remember visiting the countryside when you were young, you probably noticed when walking through a fully grown wheat field that the crop was very high, usually 4 feet tall. In fact, if you walked through these fields when you were a child, you probably wouldn't have been able to see over the top of them. But things have changed quite dramatically over the last thirty years. Now the wheat in most of the fields in the world is a strain that has been genetically engineered to be strong and not too tall, and to produce large buds of wheat germ. This strain is called 'dwarf wheat'. If you walk through a full-grown wheat field today – in England, America, Asia and most places – you will notice that the wheat is now only 2 feet high. The tall swaying wheat fields of relatively recent times have disappeared, probably never to return – and certainly never to be seen by the younger generations. Why does this matter?

First, because dwarf wheat is much more profitable for farmers: it guarantees a larger volume of wheat per acre of land at the expense of the quality of the grain, and therefore our nutrition. The story of how wheat has changed does not stop at the hybridization techniques to produce a single strong worldwide strain. The processing of wheat has dramatically changed from the days of wheat mills that were situated picturesquely along sparkling rivers. Those mills used the power of water to split and grind down the wheat. This made it easier to store, transport and cook. Now the old flour mills have been replaced with high-tech processing plants where the goodness of the wheat, the outer covering, is removed, leaving the sweet inner wheat germ. The dwarf wheat that makes up the flour present in many of our foods today, from bread (white or brown – there is not much difference) to crackers and pasta, is so highly processed that within thirty minutes of us eating it, it has turned into pure sugar in our bloodstreams (we will see the metabolic effect this has on our bodies in chapter 11). This goes part of the way to explaining why many of the patients I see, who have struggled with their weight throughout their lives, will label themselves 'breadies'. They are addicted to the reaction their body gives them to this highly processed food, just as some people are hooked on the buzz they get from sugar, and still others hooked on opiate drugs – the brain-signalling pathways are just the same.

Summary

We learned in this chapter that the way our cells process energy, and how we are reliant on those ATP micro-batteries, follows the ancient biological rules set down 4 billion years ago when cellular life first began. It wasn't until over 3.5 billion years later that evolution accelerated, when our one-celled ancestors took in and nurtured those powerful bacterial tenants (the mitochondria) using oxygen to super-charge our micro-ATP batteries. This type of energy is used by all animals, including us. These ancient energy rules are embedded deep within the layers of our evolutionary history. They ensure that energy use is budgeted for, depending on the size of the animal.

So, humans could not develop their large, energy-sapping brains without sacrificing another organ's use of their limited energy budget. We learned that prior to *Homo sapiens* developing, *Homo erectus*, our nearest relative, had tamed fire. More importantly they had started to use fire to break down the energy in food, to make it easier to digest – or, as we now call this, cooking. The energy saved by starting the digestion process of foods before they were eaten rendered our long digestive system redundant. As time passed, we evolved shorter guts and the energy saved meant that finally we could afford to develop a larger brain.

It also explains why we, as humans, are so inquisitive about food, why we love to prepare and cook and experiment with it, why it fascinates us so much. Cooking defines us.

As the millennia passed, our evolutionary instinct to control food led us to develop agriculture, and then to trade food in markets. More recently we discovered how to process food so that it could be preserved for overseas trade, becoming a business commodity. Sugar and wheat became staples in our diet.

But then, a final chapter emerged in our relationship with food. Our natural fascination with it meant that controlling its production was not enough. We wanted to understand it. In chapter 8 we learn the unfortunate consequences of this.

As we wearily pour our bowl of Frosties (almost 50 per cent sugar) in the morning, we humans know our food nirvana has arrived. Finally, we have constructed a world where our beloved sugar has found its way well and truly into our very own food chain. The feeling of calm ecstasy as our cereal goes down is our developmental pinnacle. The energy saved from preparing and cooking foods has helped us to evolve to be human. Now we have taken the next inevitable step and processed and manipulated food to make ourselves feel great. We are the Master Chef. The Master Chef has won.

But as we open our morning newspaper and sip our tea, a report catches our eye. Next to the advert for Disneyland is the headline 'SCIENTISTS DISCOVER A NEW SUPERFOOD'. Intrigued, we read on.

EIGHT

The Heart of the Matter

How Poor Nutritional Science Led to
Bad Eating Habits

'If we knew what it was we were doing, it would not be called research, would it?'

Albert Einstein

The heat was unbearable, the crowds were massive. We lurched up the picturesque street to our next destination. Families with very loud, small children surrounded and jostled us. The occasional battering ram of a pushchair made us scurry sideways. I had tried and failed to fix our trans-port problem by going to the electric buggy store, but was told I did not qualify for one: I was not 'big' enough. I watched enviously as others rode them around. We struggled on.

This was a closed community where everything was controlled by the company. The security detail herded us safely over roads, the cleaners kept the streets immaculate despite the untidy throng, and the storekeepers wore smiles of joy. People came from far and wide to be immersed in this utopia. There was no crime, no advertising, no politics. It was a dream world in which to bring your children for a few days.

The entertainment was focused on causing you to feel terrified. This may sound like a strange way to enjoy yourself, but the adrenaline and endorphin (like morphine) surge to the brain afterwards was what drew us all in. It was a safe kind of danger. When the post-terror rush wore off there was another way to get high. This was why the children flocked here. The only foods available were burgers, fries and sweet sodas. Every other store on the street was a sweetshop – laden with that wonderful food drug: sugar.

We found a bench to rest on. I needed to relax as I was scared of the terror to come; embarrassed because my eldest daughter was not afraid. I opened the large bag of pick 'n' mix sweets and felt the calmness fill me as the sugar hit my brain. My children wanted to move and live here for ever.

I looked at the families walking past us. I noticed that in this place most people were large. Only occasionally did we see a family that looked fit and healthy. A mother was laden down by both her bags and her belly fat; the father, bear-like, stabilized himself by pushing a pram. Two big live-wire kids jumped and fought; and in tow the grandma, regally corpulent in her prized electric buggy. Most of the families had sorted out their food for the day by signing up to an all-you-can-eat all-day meal deal. They clutched gigantic cups containing Coca-Cola and other 'soft' drinks. Refill stations were readily available, so their sugar rush could be continually topped up through the day by sipping through a straw.

Maybe this was the kind of world that humanity was moving towards? My reverie was interrupted by a tall and disturbing figure walking menacingly towards us. As he neared us, I recognized his familiar features – Goofy had arrived. It was time to get up and walk towards the Fairy Castle. The 'Runaway Train' could wait for now.

*

Are we slowly constructing a world like a theme park around ourselves? A world where it is difficult to access natural, fresh food? A world built on the hedonic enjoyment of sugar and fast food, but one where anxiety and stress stalk its inhabitants and that ultimately leads to a population suffering with obesity?

We know from our history lesson in the previous chapter that, as humans, we could not have evolved into the intelligent beings we are today without the help of our relationship with food and cooking. Cooking and preparing food gave us the metabolic room to shrink the size of our guts and expand the capacity of our brains. It is perhaps only natural that we would continue developing our relationship with food by discovering more and more ways of making food enjoyable. Once we opened the Pandora's box of sugar and discovered the pleasure it could give, was it not just a matter of time before we mass produced it for the whole population?

Let's take a look at the figures. If we measure sugar consumption per person since the 1820s (the golden age of healthy eating), there was a slow and inexorable rise over the next 100 years. In 1820,

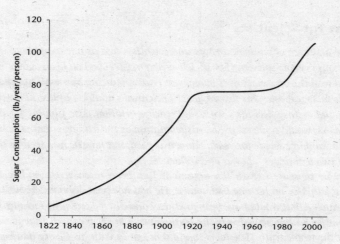

*Figure 8.1 USDA graph showing the increase in sugar consumption
from 1822 to 2000. Sugar consumption then stabilized until the 1980s*

Source: US Department of Commerce and Labor, USDA Economic Research
Service.

consumption was just 5lb per person per year; by 1920 this had
increased to 80lb per year.[1]

This coincided with the more widespread availability of sugar
once the beet variety was available as well as sugar cane. Sugar
became cheaper during this period, which encouraged food manufacturers to include it in a wide range of products. However, from
1920 sugar consumption stabilized. The Great Depression and the
Second World War would have had a significant effect on sugar
availability and affordability until the 1950s, but after this time
and for a further thirty years, sugar consumption remained stable.
It was as if we had reached a natural saturation point, or peak of
consumption.

What happened next could not have been predicted. In the 1980s,
after thirty years of consuming roughly the same amount of sugar
each year, suddenly we started to consume more and more of it again.
The price of sugar rose, doubling on average over this period, but this
did not stop an inexorable rise in our sugar consumption: from 80lb
per person each year in 1980 to 100lb in 2005. Why did we suddenly
change our eating habits again?

The Fat Scientists

The audience of eminent doctors and scientists rose as one to applaud the famous keynote speaker. He had delivered a master class on his research. He had won the argument over his biggest rival, trouncing him with indisputable facts, exposing his flawed logic. The crowd's adulation filled him with joy and ecstasy. His life's work had reached fruition. The funding for his research would come rolling in, his reputation as the leading scientist in his field would be secure for years. Fame was good, but now he had secured the top two real prizes – power and influence.

The pressure to win the argument had been intense. He had to be congratulated on how he had done it. He had not been dishonest about his research – that would have been unethical and discredited him. Technically what he had presented was the truth. But he knew very well that it was not the whole truth. The facts that did not fit in with his theory had been conveniently left out.

What he did not expect was for his research to harm people. Unfortunately, in this case it did. Sometimes there are unpredictable outcomes to seemingly well-intentioned research ideas being 'proven', particularly if those ideas turn out to be false. In the case of Dr Ancel Keys those outcomes amounted to ill health, misery and early deaths for millions of people.

In the 1950s there was a sharp rise in the incidence of heart disease in the USA. More and more people, particularly men, were succumbing to heart attacks or becoming disabled with angina. The emerging public health problem was brought starkly to the top table of government with the sudden heart attack of President Eisenhower in 1955. Scientists began to think that there might be a link between the rise in heart disease and diet. And the two main suspects in the diet were fat and sugar.

Dr John Yudkin was a British nutritional scientist who was convinced that sugar was the culprit. Since 1957 his articles and research have cited sugar as the primary cause of not just cardiovascular disease but also tooth decay, obesity and diabetes. He published a damning book on sugar entitled *Pure, White and Deadly*. In it he wrote: 'If only a small fraction of what is already known about the effects of sugar were to be revealed in relation to any other material used as a food additive, that material would promptly be banned.'[2]

His research received a large amount of interest and it looked as if his arguments were compelling enough to change the public perception of sugar. However, the sugar industry had taken note of the shifting negative publicity and decided to act first. In 1967 they donated large sums of money to three prominent Harvard scientists who conducted research that exonerated sugar and shifted the blame for heart disease firmly onto fat. The scientists were very well respected and their joint paper was published in *The New England Journal of Medicine*, the most respected US medical journal of the time.[3] The sugar money was well spent: publication of such a review in a highly respected journal could not be ignored by the medical community, and the scientists' views and opinions diffused into mainstream thinking about the dangers of fat, particularly cholesterol, to the heart.

The sugar donations were kept secret until 2017 as scientists at the time did not have to reveal who was paying them and conflicts of interest were commonplace.[4] Most of the scientists are no longer with us, but the legacy of their work was the first part of the jigsaw of evidence in what was to become known as the 'diet–heart hypothesis'. The theory was that saturated fat caused heart disease. Several more pieces of the poorly fitting jigsaw were needed before the argument was won and our nutrition was changed for generations.

Ancel Keys was an American epidemiologist with a background in high-quality nutritional research (including the Minnesota Starvation Experiment discussed in chapter 1) and had convinced himself during a sabbatical period living in England that the high-fat diet of fish and chips or Sunday roasts was the cause of high levels of heart disease in the British. He proposed that the cholesterol found in saturated or animal fats caused atherosclerosis (furring and narrowing of the blood vessels of the heart), which led to heart disease. He was another respected scientist the sugar industry was looking to for support. He did not disappoint his backers. His first attack on John Yudkin's theory was to highlight research that linked sugar intake to smoking. The research suggested that the more you smoked, the more sugary hot drinks you were likely to consume. How could Yudkin's theory hold up in view of this correlation between smoking and sugar? Ancel Keys did not hold back with his criticism of his nemesis and whenever possible tried to humiliate him and belittle his research in the scientific press or at conferences.

As part of his research, Keys published the Seven Countries study.[5] This research study looked at the relationship between heart disease and the amount of fat in the diet in seven different countries.

When the relationship was plotted on a graph, it showed a significant correlation between the two factors. This research seemed to prove beyond dispute that a high-fat diet caused heart disease. The two countries with the lowest fat intake were Italy and Japan and these countries also had the lowest levels of heart disease. The UK (England and Wales) and USA had the highest fat intake and also the highest rates of heart disease. The results when plotted on a chart were compelling; such a high degree of correlation between the amount of fat that a population ate and its level of heart disease meant there had to be a direct link.

However, what was not clear from the research paper was that Keys had originally looked at the eating habits and rates of heart disease in a total of twenty-two countries, not just seven. Ancel Keys' study had only examined countries that he thought would prove the theory; for instance, he did not study two European countries whose populations ate a lot of saturated fats but did *not* seem to suffer high rates of heart disease – France and Germany. These countries were not selected for research despite being two of the largest countries in Europe. The Dutch ate the same amount of fat as the Italians, but had double the rate of heart disease. The Swedes ate much more fat than the Australians, but the Australians had double their rate of heart disease. All the countries which did not fit into Keys' thesis that dietary fat causes heart disease were excluded.

If the paper had been 'The Twenty-Two Countries Study', it would have concluded that there was in fact no significant correlation between heart disease and saturated fats.

Shaky Foundations

Research bias by omission has been endemic in the scientific community for many years. Although, finally, this has started to be addressed, unfortunately it has left a legacy that much of our medical science is based on shaky and biased evidence. When you combine this with the influence the pharmaceutical and food industries have had over scientists, not only do you have poor research but also research that has been financed by big business and that does not always benefit those it should benefit.

Let me give you an example. Say I want to sell more of a product; it could be a drug or a type of food. I go about trying to sell more of the product by telling people that it has health benefits. Let's say I want to sell milk and I want to prove beyond scientific doubt that if you

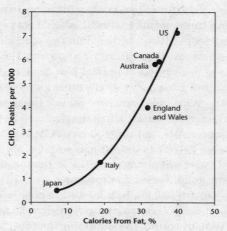

Figure 8.2A

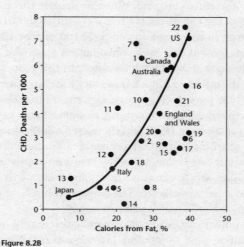

Figure 8.2B

Figure 8.2 Comparison of statistics from the Seven Countries study and fifteen other countries: Figure 8.2A shows the correlation between total fat consumption as a percentage of total calorie consumption, and mortality from coronary heart disease in seven countries; Figure 8.2B shows the same but for all twenty-two countries where data was available.

Source: J. Yerushalmy and H. Hilleboe (1957). Fat in the diet and mortality from heart disease; a methodologic note. *N Y State J Med*, 57 (14), July, 2346.

drink milk every day for five years you will be able to run faster. First, I commission a study and ask a scientist, whom I am paying, to recruit twenty young people. He splits them into two random groups of ten. He measures how fast they can run 100 metres. The scientist then tells one group to carry on as normal and come back in five years (this is the control group). He asks the other group to drink a pint of milk every day. When five years have passed, the scientist asks both groups of volunteers to come back to the track and he times how fast each group, on average, can run 100 metres. When he analyses the results, however, he gets an unfortunate surprise. My theory has not been proven! There was no improvement in the running time of the milk-drinking group. They didn't run faster.

Scientific journals will not be interested in publishing this experiment because it didn't show anything of interest. So how can I prove my theory? Well, by cunningly performing the experiment ten times instead of once – yes, ten groups, each with twenty volunteers, and ten different scientists are paid. When we analyse the results, we find that in two of the experiments there was a difference in the time the milk-drinking runners took to complete 100 metres. Unfortunately, one of them showed that milk consumption made you run slower – but the other one showed what I wanted. In this group, the subjects that drank the milk had a much faster race time after five years. I have no legal duty to publish every single experiment that I perform, especially ones that don't show anything. However, I now have one study that has an interesting result and therefore I ask the scientist to publish it in a scientific journal. The other nine studies I discard. When the publication date is due, I give my favourite journalist the tip-off that there is an article about nutrition and health that might interest him about to be published. The newspaper headline the next day reads: 'Drinking milk makes you run faster!'

This hypothetical example demonstrates that if you perform enough studies, then by random chance you will come across one whose results match what you want to prove. The milk industry, which in this case I am working for, is happy because people will be drinking gallons of milk for years to come. And the real trick? Because it is a five-year study it cannot be disproved by another lab for five whole profitable years.

Historically scientists did not have to disclose whom they were working for, or who was providing the money for their experiments. Industry, whether it be food companies or pharmaceutical companies, could control the direction of scientific research and, by perfectly legal strategies of omission or selective analysis of results, skew the outcome in their favour.

Once the direction of research is established, scientists, guided by more industry money, will follow – often on completely the wrong track. Layer upon layer of flawed research may well underpin our understanding of many areas of medicine. The more powerful the industry, or the more money it can make available, the more influence it has over scientific direction and 'scientific fact'.

Nutritional research is particularly sensitive to bias and flaws, unless it is part of a controlled experiment where the subjects can be constantly observed. Most nutritional research is epidemiological in nature, meaning that associations between the way people live and the diseases they develop are looked for. Unfortunately, these associations don't always translate into establishing genuine causes. There are often other factors involved that have not been studied. For instance, in Ancel Keys' Seven Countries study there seemed to be an association between the fat intake of a country's population and heart disease. However, he did not comment on the fact that the countries with the lowest fat intake and the lowest rate of heart disease, Japan and Italy, also had the lowest consumption of sugar, and the countries with the highest reported rates of fat intake and heart disease, the UK (England and Wales) and USA, were high sugar consumers. An independent analysis of Keys' studies several years after they had been published did, however, find a strong correlation between a particular food type and heart disease. You guessed it: sugar.

Nutritional epidemiological research, which forms the bedrock of our dietary advice on healthy eating, has many weaknesses. Eating habits are gauged by notoriously inaccurate recall questionnaires and the reporting of diseases is often based on symptoms and not actual diagnosis. If you push this 'loose data' through selective statistical analysis and throw out the results you don't like, then virtually anything can become the truth. And the 'truth' will be determined by the biggest industry backers.

The Diet–Heart Controversy

Fatty deposits in the walls of blood vessels were first described by the famous German pathologist Rudolf Virchow in the nineteenth century – he linked the deposits to heart disease. About 1 in 500 people suffer with a hereditary condition that causes cholesterol levels in the blood to be very high (called familial hypercholesterolaemia). The high cholesterol in the blood eventually causes bright-yellow fatty deposits underneath the skin of the eyelids and tendons (called xanthelasma). It was first recognized by doctors in the 1930s that people with these signs commonly died very early from heart disease (at the time it was a very rare condition). Once cholesterol could be tested in the blood – in 1934 – the first definite link between high cholesterol and heart disease was established, but this was only in the people suffering from the very rare hereditary condition. A further famous study fed rabbits (who would normally eat lettuce leaves) a high-fat diet and observed that they developed atherosclerosis (the precursor to heart problems) in their arteries. Other studies suggested that cholesterol levels in the blood could be altered in some people by changing their diet. Some people who went onto a low-fat diet lowered their blood cholesterol level.

This was the historical background to the diet–heart hypothesis debate of the 1960s and onwards. It was suggested that because high cholesterol caused heart disease (in a rare genetic condition) and that cholesterol could be lowered in some people by a low-fat diet, then by association a low-fat diet could lower a population's risk of heart disease. This seemed to be confirmed when Ancel Keys published his Seven Countries study, which seemed to strongly link dietary cholesterol and heart disease.[6]

Unfortunately, for the proponents of the hypothesis, things were not that simple. They assumed that high saturated-fat intake was to blame for the rise in heart disease in the 1950s, though in fact red-meat consumption had already been decreasing for some time.[7] In addition, their understanding that heart disease was due to a slow furring up of the arteries of the heart did not fit in with the sudden decrease in heart disease seen during the Second World War (when food, including sugar, was rationed). If the condition was chronic, then how could it be reversed so quickly? Finally, epidemiologists had failed initially to link smoking, which was at a post-war peak in the 1960s, to heart disease.

In later years, it was shown that cholesterol had many different forms in the blood depending on how it was being carried (being insoluble it needed a vehicle, called lipoprotein, to travel in the blood). One of the vehicles, high density lipoprotein (HDL) cholesterol, was very good for you and protected you from heart disease. The other, low density lipoprotein (LDL) cholesterol, was thought to be harmful. However, recently LDL has been discovered to be made up of two further subtypes: LDL type A, which is small and dense, and LDL type B, which is large and buoyant. LDL type B is *not* linked to atherosclerosis because it is too large to get into the blood vessel lining and cause inflammation. Dietary saturated fat does increase LDL cholesterol, but it's the harmless type B subtype. The small and dense LDL type A cholesterol is the cause of atherosclerosis and ultimately heart disease. The latest research suggests it is increased not by fat, or by cholesterol either, but by carbohydrate and sugar,[8] exactly what Dr John Yudkin had been saying in the 1950s before being discredited by the sugar scientists.

(For more information on the cholesterol controversy and its effect on our eating habits, see Appendix 1: Cholesterol.)

The New Science of Nutritionism

The debate on cholesterol in the 1950s and beyond still resonates in our way of life today. Just as the researchers in infectious diseases had traditionally used epidemiological studies of the environment to understand and treat infection, so nutrition scientists now looked at the food supply to the population as a potential cause of disease. This was a new way of thinking about food and disease: identifying which individual components of foods contributed to which disease.

This time, there were constellations of interested parties: politicians and their lobbyists, food manufacturers (paying the lobbyists) and their profits, the scientists and their funding (paid by the food industry), and lastly the confused consumers (who generated the profits for the food companies). The man or woman on the street now had to gauge what was best to eat in an era when food choice had gone from simple traditional seasonal fare to a confusing range of preserved, processed and imported foods. The outcome of the 'fat versus sugar' debates of this time, as we will see, has had a profound effect on the type of foods that we eat today – and an *adverse* effect on our health and waistline.

The Diet–Heart Hypothesis Becomes Official Policy

Despite Ancel Keys' study, the *New England Journal of Medicine* paper and several other epidemiological studies showing an association between saturated fat and heart disease, there was still concern in the scientific community that there was not enough evidence to make the diet–heart hypothesis public policy. Many scientists based in Britain did not share the view that the evidence was overwhelming. They looked at the evidence from more robust trials, called controlled trials. These trials compared heart disease in two groups after several years of observation. One group was put on a diet with reduced saturated fat, while the other group continued with their regular diet. Many of these experiments were carried out on thousands of people over many years. The size of the trials meant that they were quite accurate and less prone to error or bias. The outcome showed no decrease in the rate of heart disease in people placed on low-fat diets. The only consistent result seemed to be a trend in the development of cancer among those subjects eating low fat. *The Lancet*, a respected British medical journal, commenting on the diet–heart controversy at the time, stated that 'The cure should not be worse than the disease', reminding doctors and scientists of their Hippocratic oath – 'First, do no harm'.[9]

In the late 1960s, a Senate Select Committee in the US had been formed to issue guidance on nutrition. Its chairman was Senator George McGovern. The committee's original remit had been to advise government on malnutrition and ways to prevent it, but by the 1970s its sights had switched to the role of diet in disease, particularly heart disease. In 1977, after much debate with leading scientists of the time, including Ancel Keys and John Yudkin, they issued the very first national guidance on nutrition. *Dietary Goals for the United States* was a validation of the diet–heart hypothesis, despite it not being proven, and although many scientists were sceptical about the evidence it became government and national policy. The McGovern Report, as it became known, suggested cutting back on fat, particularly saturated fats containing cholesterol.

The report was a defining moment in public health. For the first time a government was advising its people what they should eat. The US dietary goals were: to increase carbohydrate consumption to 55–60 per cent of energy intake; to decrease fat consumption from 40 per cent to 30 per cent of energy intake; to decrease saturated fat to 10 per cent; decrease cholesterol consumption to 300mg/day; and to decrease sugar and salt consumption.

What happened to heart disease? Well, from 1980 to 2000 the rates of heart disease dropped from approximately 250 per 100,000 of the population to 160 per 100,000. Rates are still dropping to this day. Supporters of the low-cholesterol diet (i.e. most people who have not analysed the research) would point to the drop in heart disease as proof that it does indeed work and that it is a good public health measure. But, as in Ancel Keys' original epidemiological study, there may have been other factors in play which caused the improvement in heart health. One overlooked fact was that by the time the McGovern Report was published heart disease had already begun to drop. In 1960, it was 400 cases (per 100,000 of population); by 1970 it was down to 300 cases.

In 1964, a seminal public health event occurred. Another famous report, by the Surgeon General in the US, warned people of the real health dangers of smoking. Let's see what happened to smoking rates compared to rates of heart disease.

	Smoking Rate Cigarettes/year	Heart Disease Rate Events per 100,000
1960	4,400	400
1964	Surgeon General's Report on Smoking	
1970	4,000	300
1977	McGovern Report – Dietary Goals	
1980	3,000	250
2000	2,000	160

Table 8.1 The effects of smoking on rates of heart disease

Sources: For smoking rates: CDC (2012). National Health and Nutrition Examination Survey, 2011–2012. CDC/NCHS. For heart disease rates: C. S. Fox et al. (2004). Temporal trends in coronary heart disease mortality and sudden cardiac death from 1950 to 1999: the Framingham Heart Study. Circulation, 110 (5), August, 522–7.

It may well be that the statistics for smoking cigarettes also explain why our dietary changes coincided with improved heart health, though this was happening even before Dietary Goals was published.

But why has so much research been conducted on cholesterol? Why is the diet–heart hypothesis still important enough for many scientists to defend it? The recent funding for research in this area is not from the sugar industry but from pharmaceutical companies. Currently the biggest-grossing drugs class in the world is statins. These drugs decrease cholesterol levels in order to reduce the risk of heart attacks and grossed $35 billion worldwide in 2010. The research direction for profit is still undoubtedly cholesterol. The pharmaceutical industry would see a big drop in profit if the diet–heart hypothesis were to be disproved, and this is why they employ many of the top scientists and top labs in the world – to try and support the crumbling edifice of this fragile theory.

What Happened to Our Food after the Low-Cholesterol Advice?

The government advice spurred the food industry into action. Aware that people's food choices would change after the McGovern Report, they were quick to move to adjust their products. In fact, the dietary guidelines had given them an opportunity to market foods as *officially healthy* according to government guidelines. The slight problem was that fats, particularly saturated fats, accounted for a large proportion of the ingredients in most processed foods. Once you reduced the fat content of the food, it had an adverse effect on palatability – basically it tasted like cardboard. But they quickly came up with a solution: they started to replace the fat with sugar. Yes, the energy food that had been exonerated as a cause of heart disease!

Newly designed and 'healthy' processed foods started to line the food aisles for the discerning consumer who had recently been 'educated' by the scientists and journalists. 'Low cholesterol' and 'low fat' labels jostled for prime position. The products tasted OK as well, if suspiciously sweet. All seemed to be well in nutritional public health and it looked like scientists and politicians had made a brave and wise decision to intervene in the eating behaviour of the population.

In 1980, as the changes to the constituents of our processed foods reached the consumer, sugar consumption, which had been steady for thirty years, started to rise again. For the next twenty-five years, more and more, then even more, sugar would be consumed: to rise from 80lb per person per year to 100lb. The McGovern Report had taken effect.

Food companies not only had to reduce the total fat content of their foods (from 40 per cent to 30 per cent of total calories) but also to decrease the proportion of the fat that was of the cholesterol-containing saturated variety. The report had recommended replacing these unhealthy saturated fats with what they identified as healthy polyunsaturated oils. Fortunately, vegetable oils, containing polyun-saturates, were cheap and available. Advances in genetic engineering of rapeseed in Canada meant canola oil (Can-ola is short for Canada-oil) became a new staple alongside soya oil.

Most nutritionists will develop a fanatical gleam in their eyes when explaining the health benefits of polyunsaturated vegetable oils: low in saturated fats (the diet–heart hypothesis lives on), high in 'good fats'. We are now consuming vast amounts of these oils when only 100 years ago they were used as lantern fuel and to make candles. What exactly is this new type of food that has been integrated into our diet thanks to the McGovern Report?

If you think that vegetable oils are simply made from compressing plant seeds (rapeseed, soya, sunflower), think again. Olive oil (which is a natural and healthy monounsaturated fat) is produced in this way, the simple extraction technique going back to the days of the ancient Greek civilization. Vegetable oil production is a little more 'industrial'. You may need a degree in chemistry, or a background in petroleum engineering, to understand its production.

Seeds from vegetable plants are heated to 180°C in a steam bath and compressed to help separate the oil. The oil is then put through another bath, not steam or water this time, but a bath made of the chemical hexane (the solvent glue sniffers are addicted to) and steamed again to help more oil to be extracted. The pulp is then put into centrifuges to spin the oil away from any remaining seed residue and phosphate is added. The crude oil is then separated, but needs further refining because it smells rancid at this stage. To produce a clear odour-free oil, processes known as bleaching and deodorizing are performed on the oil. Bleaching, as the name implies, requires the use of bleaching agents to remove impurities (such as chloroform) from the oil. The deodorizing process uses scorching hot (500°C) steam at high pressure to clean the oil of bad smells.

The production of our healthy vegetable oils mirrors the produc-tion of any type of crude oil. Just as had been achieved with the innovation of sugar production, human ingenuity enabled us to manufacture a new type of food: an apparently pure healthy fat that could be added to foods and used in cooking; a foodstuff that was fit for storage and transport and trade around the world; and a food derived from previously inedible, and sometimes poisonous, plant seeds. We had done it again; 'progress' continued.

Was this really a food – or a man-made chemical that had been adapted for consumption without apparent health consequences? Vegetable oils are suspiciously like the fire-lighting oils that can give you a scare if you put too much on the barbecue. Has evolution taken us so far that our fat consumption can be sorted with substances that can be modified to run a car?

It's wishful thinking to believe that these new oils really are healthy options. Yet the consumption of vegetable oils has skyrocketed since the 1970s, helped in every decade by the cholesterol scares (see Figure 8.3). In fact, the popularity of vegetable oil has offset the reduction in animal fat and is responsible for our *total* fat intake increasing since 2000, the opposite of what the original guidelines advised.

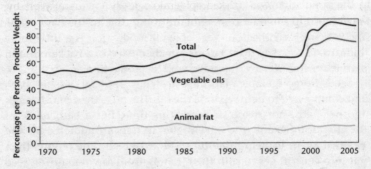

Figure 8.3 Consumption of added fats and oils increased 63 per cent between 1970 and 2005

Note: In 2000, there was a dramatic increase in the number of firms reporting vegetable oil production to the US Census Bureau.
Source: USDA, Economic Research Service, Food Availability (per capita) Data System.

Out of the Frying Pan . . .

Vegetable oils are purported to be full of the health-giving omega-3 fatty acid (we will look at the fatty acids in more detail in chapter 9). The problem with omega-3, though, as far as the food industry is concerned, is that it causes food to go off and become rancid (this is actually a sign that it *is* food and not a manufactured food replacement). This means that vegetable oils have to be treated by a process called hydrogenation to ensure they do not contain too much omega-3, which would shorten its shelf life.

Hydrogenation turns some of the omega-3 from good guys into very dangerous bad guys called trans-fats. Trans-fats cause heart disease. In fact, they are very potent poisons that once in our bodies will increase those bad-cholesterol particles – the LDL type, small dense molecules that can burrow into your arteries, causing inflammation and atherosclerosis. In addition, trans-fats decrease the amount of the good HDL cholesterol in our blood, exacerbating the risk even more (for further information, consult Appendix 1 on cholesterol).

We had therefore come full circle in our quest to reduce heart disease risk. By lowering use of animal fats and increasing vegetable oils, we inadvertently increased trans-fats and increased our risk of heart disease. The detrimental effect of this change on heart disease in the population was masked epidemiologically for many years by the decrease in smoking rates and advances in the treatment of blood pressure.

Trans-fats caused a major public health scare when it was finally revealed how dangerous they are. Governments have now encouraged food manufacturers to decrease or eliminate the amounts found in vegetable oils. However, because of the nature of vegetable oils – the fact that they will turn rancid unless treated by hydrogenation – trans-fats will always be around. Even the process of heating a vegetable oil too high in the frying pan or oven can produce these unwanted toxins. After the McGovern Report, lard and butter, the traditional stable fats (high in saturates) used in baking had been replaced by a solid form of vegetable oil called 'shortening'. Vegetable oil is liquid at room temperature, and there is only one way to make it solidify: you guessed it, more hydrogenation, meaning more trans-fats in all types of processed foods – cakes, biscuits, crackers, doughnuts, pies and margarine.

So, how much trans-fat is too much? The new US Food and Drug Administration (FDA) guidelines state that intake should not exceed 1 per cent of total calories (20 calories or 2g/day). This amount can be found in a single serving of cake, biscuits or crackers.

Hey, This Soap Looks Like Lard!

One of the most fascinating stories in our roller-coaster ride through the history of food processing is that of Procter and Gamble. William Procter, a British candle-maker, and James Gamble, an Irish soap manufacturer, found themselves related by the sisters they married – and both families settled in Cincinnati. They teamed up in business and bought the patent to a new technique from Europe that processed vegetable oils from a liquid to a solid form. Convinced that this would be a major breakthrough in soap manufacture, they set up a lab and a factory to start production. The lab produced a solid white substance which could have been used for soap but looked remarkably like . . . lard. By 1910 they had licensed their new product for human consumption. The product was the first and original hydrogenated vegetable oil, named 'Crisco'. Within a few years, even before the cholesterol scare, it became a regular household food – laden with, you guessed it – trans-fats.

Mix It All Up

Most processed food is made up of a combination of sugar and fats, mixed in with a sprinkling of salt. Quite often highly refined flour will also be added. The concoctions are finished with colours, flavourings, emulsifiers and preservatives to imitate the taste of unprocessed foods and to disguise their offensive qualities. Different consistencies of the processed foods, such as soft, chewy or crunchy, add a further level of pleasure when we eat them. Processed foods are designed in labs and tested on volunteers to see which combination hits the 'sweet spot'. The more pleasurable and addictive the food, the more it will sell. This is basic market economics: you must try and have a better product than your competitor.

A 2016 survey of the eating habits of over 9,000 US citizens showed that a massive 57 per cent of their daily calorie intake was from highly processed foods, and these foods accounted for 90 per cent of the added sugars in the diet.[10] Processed food is big, big business. Nestlé, one of the largest food companies in the world, has an annual turnover of $91 billion.

Unfortunately, the long-term health consequences of the consumption of addictive high-calorie foods do not need to be considered by the food manufacturers. They provide the product, it is up to the consumers to try and resist eating too much of it. The products sound healthy, but their labelling can be confusing, such as 'low fat' (in high-sugar products) or 'no added sugar' (in high-fat products). And the nutrition labelling is hidden on the back of the product and is suspiciously difficult to interpret. I find that having a calculator and a maths A-level is helpful, but the labels are still difficult to decipher. So, we have gorgeously tasty foods, with colourful labels saying they are healthy, all at reasonably low prices. And a consumer group of very vulnerable people. Those vulnerable people are you and me, humans, *Homo sapiens*. The same species which evolved only because it learned to cook food has now manufactured its own types of food – and we love them.

Adapting to Our New Environment

We first controlled fire, then learned how to use it to cook. We evolved large brains with the metabolic room cooking gave us. Now we have used our large brains to construct a quite unnatural food environment. Free-market economics have spread our manufactured foods around the world. Our intelligence has also allowed us to change the world in which we live in favour of one that is supposed to be more comfortable and convenient. This has translated into cities containing millions of people living together as neighbours, but often having no community. People don't need to move and undertake manual work to survive nowadays. Day- and night-time is blurred by noise and artificial light. Stress levels can be high; sleep can be difficult. There are pollutants that we are not used to all around us. Our Disney utopia may be close, but is it really what we need?

What has our new environment bestowed upon us? We now have fantastic healthcare systems that have conquered the conditions that killed our hunter-gatherer ancestors, but as our healthcare has improved so we have developed more and more 'diseases of civilization'. Our healthcare systems are designed to treat these new diseases – that are thought to occur because of the changes in our environment and living conditions. They include heart disease, high blood pressure, Type 2 diabetes, depression and cancer. One overriding condition contributes to all of them – obesity.

Common diseases and causes of death in nomadic and modern societies

Hunter-gatherer	Modern Humans	
Infection	Diabetes	
Accident	Heart disease	Obesity
Childbirth	Cancers	
Starvation	Depression	
Predators		

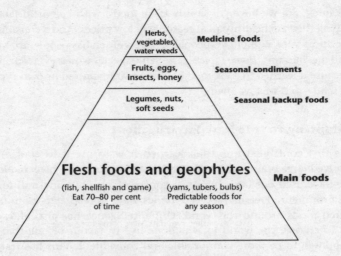

Figure 8.4 Hunter-gatherer food pyramid

Source: Adapted from M. Sisson (2012). *The Primal Blueprint*. London: Ebury Press.

Let's compare the food pyramids of these two populations.

Hunter-gatherers eat mainly meat and a staple carbohydrate. All food is natural and therefore contains an abundance of natural goodness (vitamins, minerals and phytonutrients). The organs of the animals eaten, the offal, is a chief source of natural fats – cholesterol is not avoided.

The modern food pyramid represents what the governments would like us to eat. Unfortunately, as we have learned, most of our calories come from highly processed foods. Most people are aware of the guidelines but do not really follow them.

If, as a population, we did stick to the guidelines, let's look at the food pyramid changes. Animal meat and animal fats (that lovely offal) have been downgraded from the ultimate health- and energy-giving foodstuffs that our nomadic ancestors recognized to what we are now told is a food that can cause obesity and heart disease. Meat (flesh foods) is now near the narrowing summit of the food pyramid – next to what we are also told are dangerous: eggs – and just below condiments. This is quite a relegation from occupying the base of the healthy diet pyramid of the hunter-gatherers. Instead of our staple foods being meat and tubers (sweet potato, cassava, yam,

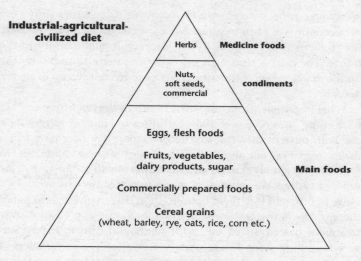

Industrial-agricultural-civilized diet

Herbs — **Medicine foods**

Nuts, soft seeds, commercial — **condiments**

Eggs, flesh foods
Fruits, vegetables, dairy products, sugar — **Main foods**
Commercially prepared foods
Cereal grains
(wheat, barley, rye, oats, rice, corn etc.)

Figure 8.5 Modern food pyramid

Source: Adapted from USDA, 1992, 'Food Guide Pyramid'.

taro, carrot, shallots, ginger etc.), the base of our new twenty-first-century pyramid is occupied by cereal grains. We are told that seeds from grasses are healthier for us and should be our new staple. The fats from meat and dairy products in our new guidelines are replaced with vegetable oils – these come in as part of 'commercially prepared foods'. Just above this, within the main foods section, is our wonderful sugar.

Hunter-gatherer Staple	New Staple
Meat, including fatty offal	Grains
Tubers	Vegetable oils/sugar

Now tell me, honestly, which diet is healthy and which causes obesity, heart disease and a whole host of the diseases of civilization?

Ouch!

Sometimes I ask my medical students how they would treat a patient who has pain in his foot because he is standing on a drawing pin? Most of them will list a range of painkillers that could be prescribed, from paracetamol to ibuprofen to codeine. Only very occasionally will a bright student give the correct answer – these are the students who I hope will have a career in public health in the future. The correct treatment is to ask the patient to stop standing on the drawing pin. If the correct advice is given, then no drugs are needed.

What has happened to our public health in the last fifty years? We have rising rates of cancer and heart disease – our two biggest killers. Both modern medicine and technological advances have had a major impact on our approach to treatment of these conditions. In fact, some would say that we are well on the way to conquering many types of cancer. We can now diagnose them earlier and we have a range of different treatment options, from surgery to targeted radiotherapy, chemotherapy or the newer immunotherapy. Heart disease treatment has advanced too, with stents and cardiac bypass surgery becoming safer. Severe obesity also affects large numbers of people and we have developed bariatric surgery and made it safer in order to reverse this condition. The whole edifice of modern medicine, it seems, is built on the treatment of these diseases of modern civilization.

But it could be argued that without the changes in our environment we would not have developed these conditions in the first place, and therefore not required these medical advances. Our expensive healthcare systems are fending off the diseases caused by the changes in our lifestyle. As a result, the life expectancy of a working-class man living in Britain in 2017 (seventy-three years) is the same as a working-class man (who had made it past the age of five years) living in the mid-Victorian era.[11] Our medical advances have thus been nullified by our new 'lifestyle diseases' in terms of life expectancy.

Vast resources are poured into research and development to beat the new diseases, but are we not overlooking the glaringly obvious treatment? Just like the patient standing on the drawing pin – prevention is better than cure. We have made great progress in curbing smoking and making it more socially unacceptable. This has had a major impact on the rates of heart disease, emphysema and lung cancer. However, other 'lifestyle diseases' are beginning to rise again as we are faced with another epidemic – obesity – leading to increasing rates of diabetes, heart disease and cancer. What should we do about this? Continue pouring more money into research and treatment, or do the sensible thing – learn our lesson from history and tackle the cause?

The New World Epidemic

Obesity is not a completely new condition, however; it has affected some people for millennia. The first sculptures of humans, dated to 30,000 years BC, depict a voluptuous woman. These clay figurines, named Venus of Willendorf after the area in northern Europe where they were first discovered, are remarkably like each other – they show an obese naked woman with very large breasts and buttocks. Any woman who was lucky enough to become so obese in these times would have been much more fertile than other women. Her buttock fat was a glaring advert for the energy reserves to carry a pregnancy through, even in times of food shortage: the perfect woman for any man looking for a successful mate to bear his children. But invariably he would be unlucky. Obesity was exceedingly rare in nomadic times, its presence probably denoting a rare genetic condition. I would guess that less than 1 per cent of people in this era developed obesity.

Once farming changed the way humans obtained their food – 20,000 years ago – there would have been a very slow rise in obesity rates, culminating in a rate of 5 per cent of the population in the mid-Victorian era – prior to the common availability of sugar. Then, as industrialization and processing and trading of food progressed, there was a further slow rise in obesity rates over the next hundred years to 15 per cent of the population by 1980.

In the 1980s there was a sudden sharp rise in the numbers of people developing obesity. The waistlines of Western populations suddenly ballooned and within a single generation obesity had become commonplace in many countries, with a quarter to a third of people affected. WHO figures from 2017 show that worldwide obesity rates tripled during this period.

The sudden acceleration in obesity rates coincided with the dietary cholesterol experiment – an experiment carried out without adequate proof that it would work, on us, the populations living in the developed (and, now, the developing) world. After the McGovern Report, our food choices which had previously been determined by our culture and family background were hijacked by the scientists. From 1980 onwards, the content of food products changed – they now contained more sugar and vegetable oils and less saturated fat.

In the fourteenth century the bubonic plague had swept across Europe, killing half of the population; the Spanish flu of 1918–19 caused the death of 50–100 million people worldwide in one year; AIDS has so far taken the lives of 25 million people. Nowadays, there are over 650 million people in the world who suffer with obesity (WHO figures for 2018). In some countries in the Middle East it is more common – if you are a woman – to be obese than *not* obese. In the future, I suspect, we will be talking in the same way about the obesity epidemic and how, in the early twenty-first century, its side effects – diabetes, heart disease and cancer – caused similar death and suffering to the wealthiest nations.

Summary

Let's recap how we got into this predicament. Why we, as intelligent humans, built this hedonic, dangerously unhealthy world that we now live in. We learned in the previous chapter how our primitive ancestors first began to change their environment and their food by harnessing the energy of fire. This gave them the metabolic room to develop larger brains. Their newly evolved intelligence continued a kinship with food. Farming, trading and processing advanced over generations until finally we saw the rise of processed food produced on an industrial scale in factories and the growth of powerful food companies. But this was not the end; our food was to change some more . . .

In this chapter, we learned how, since the 1950s, scientists have disagreed over whether sugar or saturated fat is a cause of heart disease. In the end the financial might of the sugar industry helped it to win this battle. The outcome? Natural saturated fats, particularly cholesterol, were identified as foods that are bad for us. Plenty of research sponsored by the food and pharmaceutical industries (first sugar money, then statin money) has perpetuated the diet–heart hypothesis.

The scientific battle over sugar and fat culminated in *Dietary Goals for the United States*, in 1977, the first ever government advice to a population on what to eat and what not to eat. As a result, a decrease in the amount of saturated fat, an increase in the amount of sugar and an explosion in the amount of vegetable oils consumed occurred

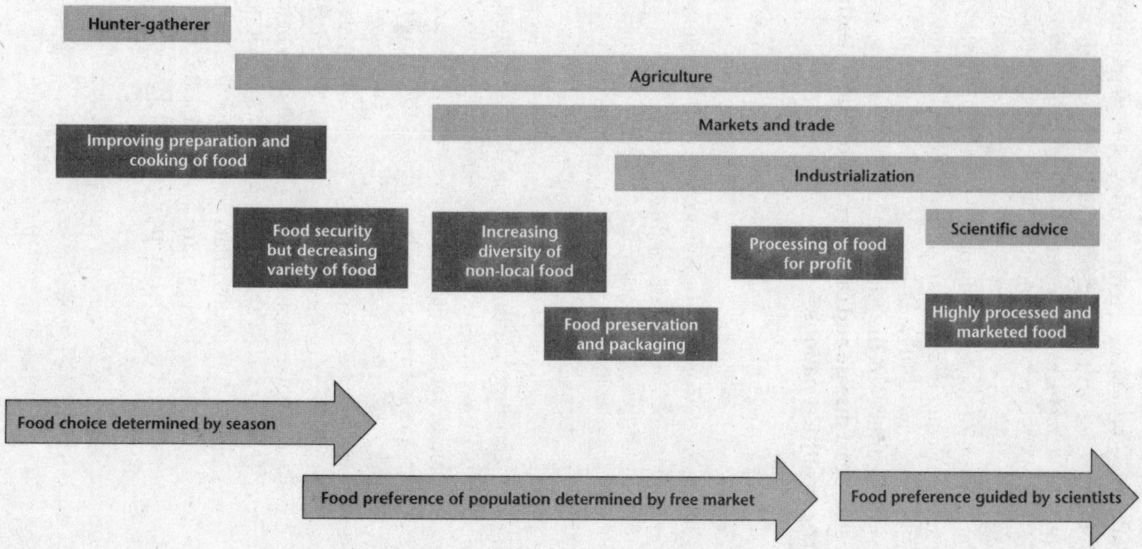

Figure 8.6 The history of humans' relationship with food

from the 1980s onwards, leading to the sudden rise in the incidence of obesity in Western populations.

This, then, is how we constructed our obesogenic environment. In the next chapters, we will see how this environment affects our health, and how best to create a new, safer food environment.

Her health was deterio...
behind the bed...
vomit bowl...
apart...

NINE

The Omega Co...

Is Obesity a Deficiency Diseas...

> Ninety per cent of the diseases known to man are caused by cheap foodstuffs. You are what you eat.
>
> Victor Lindlahr, nutritionist and
> author of *You are What You Eat*

One dark winter morning in London, just as I had reached the hospital, I was called urgently by my team to see a new patient on the ward whom they described as 'sick'. 'Sick' tends to be informal medical jargon for 'extremely unwell', so I headed quickly through the clutter of equipment blocking the hospital corridors, dodging stray wires from scanners and swerving around breakfast trolleys.

When I arrived at the bedside I was greeted by my team of efficient junior doctors, with their medical students in tow. A crowd had gathered around our new patient. I introduced myself to her and sat down on her bed, so she could tell me what the problem was. Sonia, our patient, did not look well at all. She was an Indian woman in her thirties. Her large body seemed to have been thrown on to the hospital bed from a height. One leg swung over the side of the bed, the other was bent; her arms were outstretched and the bed sheets were twisted around her as if she had been unable to make herself comfortable. She was not in a hospital gown and it looked as if the nurses had given up removing her normal attire halfway through the job. Sonia looked very tired. When I tried to talk to her, she was too weak to reply: only a whisper came out and I could not decipher it.

I looked around the bed for clues. A wheelchair indicated that her condition had been better when she had arrived the previous night, but now it didn't look as if she could even sit up in a chair at all.

...ing rapidly. A 'Get Well Soon' balloon hung ...gnalling that she had concerned relatives. Several ...were scattered around, none of them containing much ...om mouthfuls of saliva and bile. I checked her vital signs. No ...mperature, her pulse was steady and her blood pressure was OK, indicating that she probably had no infection or internal bleeding. Her breathing was normal and when I pressed on her abdomen it was not tender – there was no sign of bowel perforation or obstruction.

Sonia weighed 130kg, the senior of my team informed me. She had undergone gastric bypass surgery one month before and everything had seemed OK when she was discharged from hospital. Her husband had informed the team that she had started to vomit within a week of surgery and this had continued for three weeks. She had started to become weaker over the preceding forty-eight hours. It was a mystery: all the routine blood tests we run on patients came back normal, apart from signs of dehydration. The juniors had given her lots of intravenous fluids since her admission and this should have improved her condition if it had been purely down to fluid loss. But she was still too weak to move or talk. She just stared into space.

Sonia's mood seemed abnormally low; she did not want to communicate and just stared at us. It was as if she had become severely depressed following surgery. One of the medical students thought that maybe she had become acutely catatonic – frozen with depression. There was a suggestion that we should call in the resident psychiatrist, as there weren't any other clues as to why she was behaving in this way.

I ordered some further blood tests and asked that the team prescribe an infusion of nutrients that we would normally give to alcoholics admitted after a prolonged binge.

The next morning, I met the team again at Sonia's bed. We had a diagnosis. She was also feeling better – slightly confused by all the fuss, but sitting in the chair by the bed, reading her magazine. She had developed a disease that has remained quiescent in the world for hundreds of years. Back then, it had affected poor rice-eaters in the tropics; then it lay dormant, waiting for the right conditions to return. It was so rare nowadays that it appeared in small print in the medical students' handbooks: Sonia had beriberi.

Thiamine (vitamin B1) deficiency can catch up with you quite fast. We only have eighteen days' supply of this vitamin in our system. It tends to affect refugees, hunger-strikers and famine-sufferers.

But now, with the advent of this new type of surgery to help people lose weight, it has had a minor resurgence. Remember, after bariatric surgery the hormonal urge to eat is taken away (as seen in chapter 6) and so if a patient vomits there is no instinctive urge to replenish the system. Once the deficiency takes hold, it causes numbness, paralysis and psychiatric symptoms – a scary combination. If left undiagnosed it can eventually lead to death. Fortunately, in Sonia's case, the cocktail of vitamins we gave her included thiamine and it immediately reversed her condition. She made a quick and full recovery. Her husband and young sons were pleased to have her back.

Lessons from History – Beriberi

The resurrection of an ancient disease in our modern world, and our difficulty in dealing with it, interested me. Beriberi had not been recognized as a deficiency for many centuries – over that time millions of people lost their lives to the condition. Before the real cause of beriberi was known, the treatments given were ineffective. It was not until thiamine was isolated as an essential vitamin that it was conquered.

At each point in the history of beriberi, many years before the true cause was discovered, doctors would be convinced that their current treatment was the correct one, even when it was clearly ineffective. I wondered whether, in the future, we might look back on our current understanding of obesity and see a similar pattern to the misunderstanding and mistreatment of beriberi sufferers before vitamin B1 was discovered. Certainly, our current advice and treatment does not seem to be working to affect the obesity crisis, but doctors are still convinced that the current treatments are the correct ones.

Like obesity, beriberi did not make an appearance as a rampant disease until food started to be processed.[1] And then it only affected those populations that consumed polished rice, rice that had the outer husk (and inner germ) removed to make it easier to store and transport. Poorer villagers in South East Asia who did not have access to rice mills were protected against the disease. They consumed wild rice prepared in the traditional way by pounding it within a bowl and then sieving the fractured husk away. This rice was perfect if consumed within twenty-four hours, but it could not be stored or transported. The oil in the remaining germ of the rice would rapidly become rancid and attract mould and insect infestation. It was therefore of no use for trading, and of no use to supply large numbers of people in remote locations, such as armies on the move. Polished rice could be stored for many months and could therefore be shipped and traded. In addition to its storage advantage over wild rice, polished rice had one other desirable quality – it tasted better. But, unbeknown to those consuming the polished rice, the husk and germ layers that had been removed contained an element essential to their health – vitamin B1. If polished rice made up the bulk of their dietary intake, which was the case for many populations, then they would be vulnerable to developing beriberi.

Beriberi was first described in ancient Chinese manuscripts dating back to 2000 BC. The name originates from a Sinhalese translation of 'weak-weak'. Roman legionaries noted that when an outbreak occurred, it could kill 30 per cent of an army and represented a greater danger than any enemy. It seemed to affect groups of people living closely together, particularly soldiers, sailors and prisoners. Early observers of the condition noted that it struck the rice-eating towns and cities in the south of China, but seemed to spare the wheat-eating peoples who inhabited the north. A dietary deficiency was therefore suspected.

However, this information was not available to British colonial doctors and scientists who were sent to analyse the condition that was killing large numbers of the King's subjects in the Far East. They concluded, at various stages in history, that the condition could be due to:

1. A miasma, or foul-smelling cloud of gas
2. An infectious agent
3. An anti-toxin contained within rice.

Misunderstanding and Confusion

The colonial scientists and researchers initially thought that beriberi was caused by a miasma, a cloud of foul-smelling air emanating from unsanitary conditions and rotting food. Even our beloved Florence Nightingale espoused this theory and was responsible for improving the air quality in the military hospitals that she ran. The side effect of cleaning up the hospitals, so that they smelled better, was that communicable diseases were better controlled and fewer people died – reinforcing the belief in miasma as a cause of many diseases.

Outbreaks of the disease affected isolated towns and villages and therefore some scientists concluded that it might be caused by an infectious agent, or a toxin. Others noted that it affected rice-eating areas only and so thought it might be caused by some sort of anti-vitamin contained in the rice. As with many conditions prevalent at this time, poor scientific communication between countries and cultures led to incorrect theories being propagated for many years by the doctors.

Even *within* countries there was a lack of consensus and understanding which cost many people's lives. In 1895, Kanehiro Takaki, a doctor commissioned in the Japanese navy, thought that protein deficiency was the cause of beriberi. He introduced extra protein rations (which happened to contain enough vitamin B1) to all sailors on long voyages and observed the complete eradication of the condition. Unfortunately, his medical counterparts in the Japanese army did not believe his theory and continued to think that beriberi outbreaks were due to an infectious disease. They continued with the same rations of plentiful polished white rice, and not much else, and tightened up on sanitation. The result? Ten years after beriberi had been completely eradicated in the Japanese navy, 80,000 of their army developed the condition, 8,000 of them losing their lives to it in the 1904–5 war with Russia.[2]

In hindsight, there was ample evidence that beriberi was caused by some sort of a nutritional deficiency. The problem was that no scientist ever had all the evidence together in one place at the same time: it was historical, disparate and scattered. The breakthrough as to the real cause, when it did finally come, was down to our old scientific friend and companion – luck.

The Breakthrough

In the 1890s Dutch scientists in the East Indies (now Indonesia) had been given the task of finding an infectious cause for beriberi. They took blood from caged hens that had developed it and injected the serum into unaffected caged hens. The hens that had been injected with the serum promptly developed symptoms of the condition – confirming their theory that beriberi was due to an infection. However, the scientists, being sensible, wanted to double-check their findings. They repeated the experiment, but this time the hens which had been injected with 'infected' beriberi blood remained healthy. Now they were confused. Why would the same experiment yield two opposing outcomes? They scratched their heads to analyse if there was any difference between the two experiments. The only disparity they could find was that the warden who looked after the hen coops had changed. When they questioned him, they found that the old warden had fed his chickens with polished rice whereas the new keeper fed them wild rice! They had finally proved that beriberi was caused by something lacking in polished rice. Within a few years, research teams had isolated vitamin B1 in the husk of the wild rice and purified it for treatment. Beriberi, after causing millions of deaths, had finally been understood and cured.

Scurvy, Another Deficiency Disease

Let's look at another example of a disease caused by a deficiency – one that was also misunderstood for centuries and caused suffering and countless deaths. This condition was thought to be due to:

1. Poor morals and filth
2. Homesickness
3. Too little exercise
4. A miasma of foul-smelling putrid air (again).

This condition could cause personality changes, extreme tiredness and cravings for foods (not unlike the symptoms experienced by dieters). Swollen and rotting gums and skin rashes would develop and wounds wouldn't heal. Eventually victims would succumb to blindness, psychosis and internal bleeding, with terrifying bouts of vomiting and the passing of foul-smelling blood.

Scurvy, caused by a lack of vitamin C, was a well-known condition even before the era of long-distance seafaring.[3] 'Land scurvy' was a common affliction of the Crusaders on their treks through the deserts of the Middle East. Napoleon's armies were similarly affected on their long campaigns, and it was noted by Napoleon's own physician that eating horse meat seemed to protect his men from the condition. The meat was fresh and contained adequate amounts of vitamin C to prevent scurvy developing. The Napoleonic soldiers developed a taste for healthy horse meat and continued to eat it even after they left the army, starting a French tradition that has been passed through the generations and continues to this day.

The British navy had noted that scurvy killed more sailors than any enemy could. Of the 184,000 British sailors to have taken part in the Seven Years War against France and Spain in 1756, over 133,000 were reported missing or died from disease, and by far the most common disease was scurvy. William Clowes, naval surgeon, wrote that 'their gums were rotten even to the very roots of their teeth, and their cheeks hard and swollen . . . they were full of aches and pains, with many blemishes and reddish stains or spots'.

The Cure – Elixir of Vitriol?

The cause of scurvy – a lack of fresh fruit and vegetables (containing vitamin C) – had been known to seafarers for centuries, but the medical establishment had never accepted this. Instead, they treated the condition with fizzy drinks designed to activate the digestive system. These drinks consisted of an 'elixir of vitriol' – sulphuric acid mixed with barley water, with spices added to disguise the foul taste. Stores of these drinks were aboard every ship in the British navy for countless years, despite having no effect as a cure for scurvy.

The real cure had been found and then lost many times over the centuries. All sailors knew instinctively that consuming fresh fruit and vegetables protected against the condition. In the sixteenth century the Portuguese had planted orange and lemon groves next to their ports so that sick sailors could quickly regain their health.

James Lind, a young Scottish naval surgeon, had gained experience of scurvy at first hand whilst on voyages to the Caribbean. His observations gave him a wider perspective and understanding of the condition than many in the medical establishment. He also had a healthy disregard for authority and a disdain for the corruption of medical knowledge. Convinced of the true cause of scurvy, he was frustrated that this was not being understood by the self-proclaimed experts.

James Lind's Trials

In 1747, Lind conducted the first ever controlled clinical trials to assess the treatment of scurvy. Twelve sailors who had developed symptoms of scurvy were split randomly into pairs and each pair was given a different treatment. The treatments on offer were:

1. Elixir of vitriol (the standard treatment of the time)
2. A quart of cider a day
3. Vinegar – two spoonfuls, three times a day

4. A paste of garlic, horseradish, balsam and mustard seeds
5. Sea water! – half a pint a day
6. Two oranges and one lemon a day.

Within days the two sailors receiving the citrus fruits had recovered and were able to return to their duties. In 1753, Lind published his *Treatise of the Scurvy* outlining his experiment.[4] In the same publication, he wrote on the state of medical knowledge, 'theories were invented . . . according to the whim of each author, and the philosophy then in fashion . . . The learned ignorance of the age lay concealed under a veil of unmeaning, unintelligible jargon.' He was proved correct . . . it took another forty years before his theories were put into practice by the Admiralty.

The British Navy, although slow to implement Lind's theories, remained ahead of all their rivals in other countries and gained a significant military advantage. In fact, the 1804 blockade of Napoleonic ships within French ports, an exercise requiring sailors to stay for months on board their ships, would not have been possible without the Royal Navy's purchase of 50,000 gallons of lemon juice for its sailors. The blockade prevented Napoleon's grand plan of a sea invasion of Britain and changed the course of history.

In 1867 all naval vessels were ordered by Parliament to carry a supply of limes and lime juice. The association of British sailors with limes spawned the nickname 'limeys' – still in use today to describe British expats living abroad – and is a reminder of a long-forgotten disease.

Could a Dietary Deficiency Lead to Obesity?

To return to the modern day, could a dietary deficiency be a signal to our body to move our weight set-point upwards? Could a nutritional defect in our Western diet be misinterpreted by our bodies as a sign of an impending famine or a long hard winter – the signal to gain some extra fat as insurance? If this is the case, if obesity could have been triggered by a deficiency in some people, it would make it similar to the historical diseases beriberi and scurvy – those deficiency diseases that were misunderstood by doctors and scientists for so long.

A deficiency, as a driver of obesity, would fit in with the weight set-point theory that we discussed in chapter 1. To recap, the weight set-point theory says that the subconscious brain (the hypothalamus) is responsible for calculating the weight that is best for us. It uses data from our genes and environment to calculate our own individual weight set-point. A change in the environment, i.e. a deficiency of an essential food, could turn the set-point upwards. Once the weight set-point is elevated (above the actual weight of the person), the subconscious brain drives the weight towards the new set-point by strong appetite signals (by giving you a voracious hunger) and a reduction in metabolism (making you tired and listless). Next thing you know . . . you are hungry all the time, can't stop eating and your weight is on the rise.

This would be a completely left-field way of looking at obesity. The voracious hunger, coupled with listlessness, would no longer be seen as the *cause* of the disease, but as a *symptom* of it – just as the tiredness of scurvy sufferers was recognized as a symptom of the condition, and not the cause of it, once vitamin C had been discovered. History would be repeating itself. But more importantly we would have an effective strategy to treat obesity.

What's Missing?
Let's go on a detective hunt for our deficiency. We need to answer three questions:

1. When did obesity rates start to rise?
2. What happened to our food at this time?
3. What was removed or replaced?

The first question is easy to answer: obesity rates started to increase dramatically in the mid-1980s. What happened to our food at this time? Government healthy-eating guidelines happened.

In 1977 the McGovern Report, *Dietary Goals for the United States* (discussed in chapter 8), was released to a trusting public. The guidelines were produced in response to a dramatic rise in heart disease following the war. The diet–heart hypothesis, despite many scientists disagreeing with it, underpinned the dietary guidelines. Saturated fat was demonized as the root cause of the heart disease epidemic (and still hasn't recovered, see Appendix 1). So the obesity epidemic coincided with a big change in the advice to the population on what to eat. If our deficiency theory holds true, then some essential element of the diet must have been removed or replaced at this time.

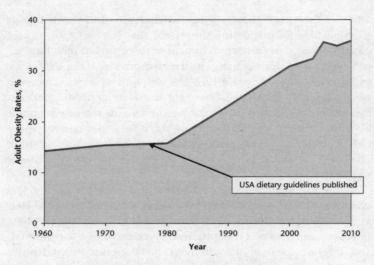

Figure 9.1 US obesity rates, 1960–2008

Source: C. L. Ogden and M. D. Carroll (2008). Prevalence of Overweight, Obesity, and Extreme Obesity among Adults: United States, Trends 1960–1962 through 2007–2008. *National Health and Nutrition Examination Survey (NHANES)*, June. National Center for Health Statistics.

'Healthy' Vegetable Oils Take Over

One of the most striking changes in eating behaviour that occurred was the replacement of saturated fats (found in butter and lard) with vegetable oils such as cotton seed, safflower, rapeseed (canola) and sunflower oil. Vegetable oils had also been hailed as protective of the heart: they had been shown to decrease the level of blood cholesterol, and this, it was assumed (if you went along with the diet–heart hypothesis), would decrease the risk of heart disease. Vegetable oil consumption skyrocketed from 15lb/year in 1970 to over 60lb/year in 2009, a rise of 300 per cent (see Figure 9.2).

The 'safer' alternative to butter, a semi-solid mix of vegetable oils called margarine, remained popular. Flora, sponsors of the London Marathon, became the symbol of healthy man-made foods.

The change in the type of fats that were consumed had a downside, however. Even though the amount of saturated fat consumed continued to decline, the increased intake of vegetable oils, shortenings (a solid form of vegetable oil used for baking) and margarine

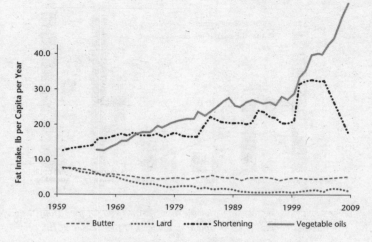

Figure 9.2 US added fat intake per person, 1960–2009

Source: Data from the USDA Economic Research Service.

actually led to a rise in the total amount of fat consumed: a rise of 63 per cent from 1970 to 2005.

A Gift to the Wheat and Corn Farmers

The dietary goals document also recommended a second critical change: the increased consumption of grains, which were considered healthy for the heart. The rise in the amount of grains consumed suited the US Department of Agriculture, which had vast reserves of wheat and corn to sell. The Western public duly followed this advice, resulting in an increase in wheat flour consumption from 115lb to 150lb per person per year from 1980 to 2000. The guidelines had envisaged people consuming whole grains, but the vast majority of wheat consumed was highly refined, and therefore had a similar effect on people's insulin levels as table sugar. Insulin levels, as we discuss more fully later in this book, also play an important role in determining our weight set-point.

The dietary guidelines did *not* stimulate a public interest in home cooking; on the contrary, people were increasingly confused by what to eat, and as a result more and more processed food was sold – quite often with reassuring labels, such as 'low cholesterol' or 'heart healthy'. Most of the dramatic rise in the consumption of both vegetable

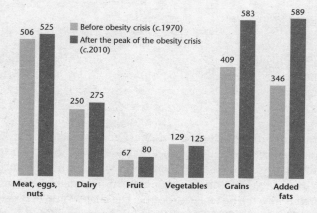

Figure 9.3 Daily calories consumed by the average American, by food group, before the obesity crisis (c.1970) and after the peak of the crisis (c.2010)

Source: Statistics from the USDA Economic Research Service, Pew Research Center, Washington DC, US.

oils and refined grains was in the form of these processed foods. The cheap oil and grain commodities could be mixed with whatever other flavourings, preservatives and e-additives were required to make biscuits, crackers, cakes, soups and gravies etc.

Food Changes Just before the Obesity Epidemic
So, in summary, the changes in our diets just before the spike in obesity that started in the 1980s were:

- Increased vegetable oil
- Increased grains
- Increased processed foods.

When you glance at the figure above, the major changes in our eating patterns become clear. Meats, eggs, dairy, fruit and vegetable consumption were all pretty similar. The rise in sugar consumption was small (30kcal). But consumption of grains rose by 170kcal/day and added fats from vegetable oils by a massive 240kcal/day.

What is Missing?

Our final question on the hunt for our deficiency – what was taken out of the food we consumed as a result of these changes? Did the dramatic rise in vegetable oil consumption, combined with food-processing techniques, somehow cause a micro-deficiency of a food essential for our health? Recent research into lipids (fats) suggests that this may indeed be the case, and that these changes may not only be contributing to obesity but also to other common Western diseases such as heart disease, autoimmune conditions and cancer. Our scientific understanding of the function of fats within our bodies lags far behind our understanding of vitamins. This fascinating new area of research is now trying to piece together the relationship between the type of fat consumed in our diets and the development of Western diseases, including obesity.

To help us understand the potential lipid (fat) deficiency that may be contributing to the obesity crisis, let's take a refresher in fats and what they do within our bodies.

Apart from their energy-storing capacity, fats have many other vital functions. Our brains and nerves are predominantly made up of fats; in fact the brain consists of 50 per cent cholesterol, so fat is essential for normal nerve and brain function. Our hormones, which are powerful drivers of behaviour, are also made from fats. These include our sex hormones (oestrogen and testosterone) and stress hormones (cortisol). Our inflammatory processes, which coordinate the repair of tissues and fight against infections, are driven by messengers derived from fat. Finally, and probably most importantly, fat makes up the cell walls of every organism on Earth, and cell walls act as the final barrier to, and conduit from, the outside world to our very core, our DNA.

Into the Frying Pan

There are three different types of fats. Each fat molecule is made up of a chain of carbon atoms.

Each of the carbon-to-carbon bonds in the chain holds the precious energy contained within the fat. One end of the carbon chain is attracted to fat – this is called the omega end. The other end of the carbon chain is attracted to water – this is called the alpha end. This configuration, the carbon chain having one fat-loving end and one water-loving end, is called a fatty acid: this is how fat exists within the body. Think of the fat molecule as a long dining table in a medieval

Figure 9.4 Composition of fat molecules

banquet, with the king and the queen (with their different attractions) sitting at either end.

The number of 'guests' sitting at the dining table determines the type of fatty acid that it will become. In fatty acids, the 'guests' at the table are hydrogen atoms. If the dining table is full of 'guests', and there is no room for any more, then it is called a saturated fatty acid. These are quite rigid and unbending. They are also very stable and will stack on top of each other and easily make solid structures.

Saturated fats, because of this stability, are solid at room temperature. Examples of foods containing a lot of saturated fat are: butter, lard, cheese, palm oil, coconut oil and animal fats. When the dining table is full, apart from a single seat that is left free, the fat is called a *monounsaturated fat* (mono = one). This type of fat is slightly more flexible than the rigid chains of saturated fat. Because of this it is liquid at room temperature but will solidify when in the fridge. Examples of foods containing this type of fat are: olive, peanut and avocado oils.

When the dining table has several seats still available, the fat is called a *polyunsaturated fatty acid*. These are much more bendable and flexible than saturated or monounsaturated fats and are therefore liquid at room temperature and also in the fridge (think cooking oil).

The Special Fats

There are two special types of polyunsaturated fatty acids. These are called omega-3 and omega-6 fatty acids and are different to all other types of fat.

Figure 9.5 Composition of a saturated fatty acid

The other fats, the saturated or monounsaturated ones, can be made by us, within our own bodies. We are not reliant upon them in our diets. As we know, the saturated fat cholesterol is an essential component of our brains and our cell walls, and therefore essential to our health. If we don't eat foods containing cholesterol (as some dieticians might advise) our bodies will take over and produce it from scratch within the liver.

The omega fats are unique: we are unable to make these fats ourselves. Just as for vitamins, we are reliant on eating foods that contain them. They are therefore known as essential fatty acids – because it is vital to our wellbeing that they are part of our diet.

Omega-3 and 6 may seem similar, but before we go any further I want to highlight some important differences in them so that you can understand how each of these fats affects your body.[5] First, omega-3 has a much curlier and more flexible carbon tail and moves faster than omega-6, changing shape many times per second. It therefore makes any tissue that contains it much more flexible, faster and more adaptable. This is a very important trait of omega-3 within our bodies. Secondly, omega-3 gets oxidized much more quickly than omega-6. This means that when exposed to oxygen it will break down, or decompose, more easily. Think of what happens to food when it is left out and unattended: it goes brown and decomposes – this is oxidation. Fresh foods that go rancid quickly, if left out, tend to have high levels of omega-3 (e.g. fish).

So, we have identified these two types of fat that, like vitamins,

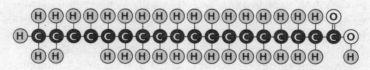

Omega-3 fatty acid

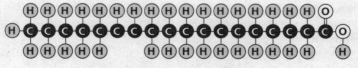

Omega-6 fatty acid

Figure 9.6 Composition of omega-3 and omega-6 fatty acids

	OMEGA-3	OMEGA-6
Carbon tail	Curly, dynamic, fast-moving	Slower and stiffer
Makes tissues	More flexible, adaptable	Less flexible, adaptable
Oxidation	Easily decomposes	More stable

Table 9.1 Characteristics of omega-3 and omega-6

are essential in our diet to maintain health. Could the changes to the types of fat we now consume – the changes brought about by the dietary guidelines of 1977 and coinciding with the start of the obesity crisis – have led to a deficiency of these essential fatty acids? Let's examine where they are produced in nature.

The Sunshine Fat – Omega-3

As the sun shines on our rainforests, meadows and seas, a process essential to all life on Earth is taking place. Embedded within the cells of all green leaves, and within all plankton and algae floating on the sea, are structures called chloroplasts. Chloroplasts are the plant equivalent of our own cellular energy factories (our mitochondria). However, they can be considered the most important structures on Earth. They function to take the energy from sunlight and convert it into chemical energy. This energy is used to produce more and more complex fats, proteins and carbohydrates so that the plant or plankton can grow and survive. The precious energy that these structures produce forms the basis of the food supply to every other creature on Earth, from the cattle and fish eating them to larger predators – including us.

All the biological energy on Earth stems from chloroplasts. But they produce something else that is essential to us as well. Chloroplasts produce omega-3. And because the world has a lot of rainforests, vast meadows and grasslands and abundant algae floating on our oceans, omega-3 is the most common fat in the world. You would never have thought that a serving of spinach or lettuce contains a fat that is so abundant, and so essential to our health.

Omega-3 is passed through the food chain, so that any animal or fish that ingests green plant matter will therefore also contain omega-3 within its cells. Fish, which don't have much food choice except plankton, therefore contain large amounts of omega-3. Any type of animal that grazes on grasses, such as sheep or cows, will also contain high levels of omega-3 in its body. A predator that eats the animal that ate the green leaves or grasses will integrate the omega-3 into its tissues as well. At the top of the food chain are human beings, and as long as we consume lots of omega-3-containing vegetables or the fish or cattle that have themselves fed on greens we will also have abundant omega-3 within our own bodies.

The Autumn Fat – Omega-6

Omega-6, the other essential fatty acid, is also made by plants, but it appears in their seeds and not in their green leaves. As with omega-3 it is passed through the food chain, so it will be abundant in animals that consume seeds – and in the animals that prey on the animals that consume the seeds.

So how did the changes in the type of foods that we were eating at the beginning of the obesity crisis affect the amount of these essential fatty acids in our diet?

The Verdict – Rapid Rise in Omega-6

To recap, the American dietary guidelines recommended decreasing the amount of saturated fat consumed and increasing the amount of grains in the diet. Much of the saturated fat was replaced by vegetable oils; in fact, overall the amount of fat consumed increased. The vegetable oils that replaced saturated fat are made from seeds and therefore contain abundant omega-6.

Soybean oil now accounts for 50 per cent of the vegetable oils consumed in the USA. They contain 54 per cent omega-6 fat by volume; that is 120kcal per tablespoon. This is now the most common oil that is added to processed foods.

The bar chart overleaf confirms the considerable quantities of omega-6 fats present in most vegetable oils compared to omega-3. The exceptions are cod liver oil (originating from plankton-eating fish) and butter, which has mostly natural traditional saturated fats and low levels of polyunsaturated fats.

The dietary advice to increase the amount of grains (seeds) eaten would also have increased the amount of omega-6 in our diets. What about processed foods? We know that processed foods contain lots of vegetable oils and refined grains like wheat, so again omega-6 will be plentiful. Far from causing a deficiency of this fat, the dietary guidelines led to an unprecedented increase in the level of omega-6

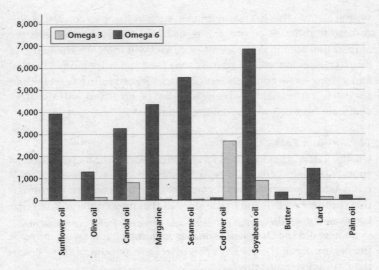

Figure 9.7 Omega-3 (light grey) and omega-6 (dark grey) and levels of common cooking oils and spreads. Units mg per tablespoon (14g)

Source: Data courtesy of USDA National Nutrient Database for Standard Reference: Nutrition Data; https://nutritiondata.self.com.

in our bodies. To those who still adhere to the diet–heart hypothesis, and remember the effect omega-6 has in decreasing the cholesterol level in the blood, this would seem like a helpful change to our diet. Let's test it . . .

A 2013 study, published in the esteemed *British Medical Journal*, looked at what would happen to our health if we replaced saturated fats with omega-6 fats. The rationale of the study was that populations living in the Western world have been encouraged to make this dietary change, even though it has not been robustly tested. The study compared two groups, each containing about 220 men who had recent heart problems. One group was to continue eating their normal diet containing saturated fats, and the other was to replace these fats with linoleic acid – safflower (a type of seed) oil and margarine (omega-6). The outcome? 'Substituting dietary linoleic acid in place of saturated fats increased the rates of death from all causes, including heart disease.'[6] Despite the clear conclusions of this study (and many similar ones), the dietary advice we receive remains the same. The NHS still advocates omega-6 vegetable oils in place of saturated fats. The

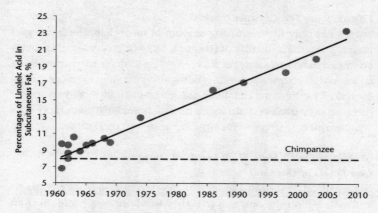

Figure 9.8 Increasing levels of linoleic acid (omega-6) found in body fat in American citizens, 1961–2008

Source: S. Guyenet (2011), Seed oils and body fatness – A problematic revisit. Whole Health Source, 21 August.

diet–heart hypothesis remains entrenched at the centre of this advice, seemingly unmovable in the face of mounting evidence against it (see Appendix 1 on cholesterol at the end of the book for more detail).

As the amount of omega-6 in our diets has skyrocketed, so, as expected, has the amount of omega-6 that we carry in our tissues. Dr Stephan Guyenet, author of *The Hungry Brain*, published this graph to demonstrate the phenomenon. Levels have gone from 8 per cent – matching other primates (chimps) in 1961 – in a slow progressive rise as more and more vegetable oils and grains have been consumed, to 23 per cent in 2008.

Omega-3 and the Western Diet

If omega-6 levels, in our post-1980s diet, look plentiful, how did the changes affect omega-3? The dietary guidelines recommended a decrease in saturated-fat intake. If a population cuts down on grass-fed meats and dairy products in order to decrease their cholesterol intake, they will also cut out an important source of omega-3. We know that any animal that feeds on grass has good levels of omega-3 in their tissues, as well as in their milk. In this respect, decreasing the amount of red meat and dairy products consumed also decreased total amounts of omega-3 taken in.

Livestock are Fed Cheaper Grains

But it's not only the decreasing amount of meat that affects omega-3 intake, it is also the quality of the meat. More intensive farming methods mean that most cattle are now fed grains to make them grow faster (as we saw in chapter 2). Seeds (grains) also contain a lot more energy compared to the same weight of grasses and can be stored for longer (because they don't contain much omega-3), making them more convenient and cost-effective for large farms to feed their cattle.

Grain-Fed Cattle Have High Levels of Omega-6 and Low Levels of Omega-3

Grain-fed cattle take in the large doses of omega-6 from the grains and miss out on the omega-3 that they would normally obtain from grass. This change in their diet is reflected in the nutritional quality of their meat (less omega-3 and more omega-6). Fish are not immune to this change in their nutritional quality either. Most of the fish now available in supermarkets originate from fish farms, and just like cattle (and humans) fish get bigger if they are fed grains rather than their natural food (plankton). At the pinnacle of the food chain are humans, consuming affordable grain-fed meat or fish – and switching the levels of key *omega fatty acids* in our own tissues from omega-3 to omega-6.

If It Has a Shelf Life, It Doesn't Contain Omega-3

How about the increasing amounts of vegetable oil and processed food that we consume – how does this affect our omega-3 intake? Remember the food left unattended, the food going brown? The oxidized food, which contained omega-3?

Processed foods, the food made in factories (and not farms), need to have a long shelf life. But, remember, any type of food that has a reasonable shelf life will have had most of its omega-3 removed. Fresh foods contain omega-3, that's why they go off rapidly when left out of the fridge. The same rule applies to vegetable oils (but not olive oil): they need to be rid of their omega-3, otherwise they become rancid quite quickly. Therefore they are chemically and heat treated to eliminate it. As mentioned previously, the hydrogenation process to remove the oxidizing potential of unsaturated fats causes the production of cardio-toxic trans-fats – all in the interest of making the food taste fresher for longer, and ultimately with the aim of making the food company a larger profit.

So the changes in our food that were supposed to help us to decrease the level of saturated fats in our diet actually resulted in a big divergence in the proportions of the two essential fatty acids consumed. The amount of omega-6 increased dramatically and the amount of omega-3 decreased rapidly.

Omega-3 to Omega-6 Ratio Changes

Ideally, and throughout our history, the ratio of omega-3 to omega-6 within our bodies would have been between 1:1 and 1:4 (i.e. four times more omega-6 compared to omega-3). If we go back to hunter-gatherer times, when all foods were fresh and the diet was not based on grains or vegetable oils, we would see this range. People living in remote areas of the world today, who consume natural home-grown foods, will have these levels as well. But if you consume a Western diet, as we have seen, many omega-3s have been removed and large amounts of omega-6 have been added to foods that have been processed or commercially produced. So, the omega-3 to omega-6 ratio rises to a staggering 1:50 in some Westernized cities.

Table 9.2 summarizes the changes in the omega-3 to omega-6 ratio over time, and the differences between geographical regions, from a 2004 study from the Center for Genetics, Nutrition and Health in Washington, DC.

How Does the Change Affect Our Bodies?

So, we have unearthed a new, modern deficiency, one that until recently had not been known about. This deficiency of an essential fatty acid

Population	ω6:ω3
Palaeolithic	0.79
Greece prior to 1960	1.00–2.00
Current Japan	4.00
Current India, rural	5–6.1
Current United Kingdom and northern Europe	15.00
Current United States	16.74
Current India, urban	38–54

Table 9.2 Omega-6 to omega-3 ratios in various populations

Source: A. P. Simopoulos (2004). Omega-6/omega-3 essential fatty acid ratio and chronic diseases. *Food Reviews International*, 20 (1), 77–90.

is made worse by food processing, and by a lack of fresh foods. The parallels with vitamin B1 and vitamin C deficiency are starting to appear. But, just as these vitamin deficiencies were revealed as the true causes of beriberi and scurvy, will the new deficiency we have identified help us understand the modern disease of obesity?

The Story of the Omega Brothers

We learned that the omegas originate from different parts of the plants: omega-3 from their green leaves, and omega-6 from their seeds. When we look at the function of the omegas within our bodies we can also see that they have many opposing effects. They are like two brothers, originating from the same mother, but having opposite personalities. The omega-3 brother is quick and fast and flexible, with a personality that is healing, but a constitution that is brittle. The stiffer omega-6 brother is much more solid and stable, but is slow and tends to cause trouble wherever he is. As with many brothers, the omega boys have major rivalries. They are constantly competing and fighting with each other for their favourite place: they want to sit on the same wall. That wall is our cell wall, a critical area for our health.

Imagine the small, friendly, quick and flexible (omega-3) brother wearing green (from the leaves he originated from) and the bigger, more stable and more angry brother (omega-6) wearing brown (from the seeds he came from). Now imagine that there are many pairs of omega brothers sitting on your garden wall, guarding it. In normal circumstances, there would be an even number of green- and brown-shirted brothers (a 1:1 ratio). If you wanted something passed over the wall to the next garden there would be plenty of friendly, flexible greens to help pass things backwards and forwards. If, one day, scores of brown shirts turned up and there were hardly any – only two or three – green-shirted brothers, the wall would be guarded predominantly by the stiff and sulky brown shirts. It would be difficult to get them to agree to pass things over the wall to the neighbour and, in addition, when you ventured too close to them you might get kicked, or, even worse, ambushed and injured, by them.

Now imagine those omega brothers miniaturized and guarding your cell walls, helping to control what comes into and out of the cell and also defending the cell against danger. This is the function of the omega brothers: they play a critical role in the functioning of our cell walls and hold the keys to entry into our cells and the weapons to defend them.

As we mentioned, the omega brothers are in a constant fight with each other to sit on our cell walls: there is limited space and some will be unable to find room. If there is more omega-3 circulating, there will be more in the wall, and the same goes for omega-6. The proportion of the omegas – 3 or 6 – within our cell walls mirrors the proportion of omega-3s to omega-6s that we eat in our diets.

The Food We Eat Literally Imprints Itself into Our Cell Walls

We know that the amount of omega-3 in our diet has fallen dramatically in the last forty years and the amount of omega-6 in our diet has increased even more dramatically. This change in the ratio of omegas in our diets is reflected in a similar change to the ratio of omega brothers within each and every one of our cell walls – all 30 trillion of them. Suddenly our cell walls have a ratio of around 20 stiff, unfriendly omega-6 brothers to every 1 quick, friendly, flexible omega-3 brother. Let's guess what this may be doing to our health.

The Functions of the Omega Fatty Acids

The (opposing) functions of the omega brothers span three main areas:

1. Defence (inflammation)
2. Cell wall permeability (insulin sensitivity)
3. Messaging (mood and appetite).

Defence

Omega-3 and omega-6 have opposing actions in their inflammatory response to infection or injury.

Omega-6 fatty acids are broken down in the cell membrane and release factors that encourage:

- Increased inflammatory response
- Increased blood coagulability (clotting).

Omega-3 fatty acids have an opposing response when released from cell membranes. They are much less inflammatory and decrease blood coagulability. So an upsurge in the cell wall of omega-6 compared to omega-3 will increase the amount of inflammation to any given stimulus: it will make our immune systems more sensitive.

If the omega-6 to omega-3 ratio goes up dramatically, it could make our immune systems hypersensitive.

A hypersensitive immune system can lead to the development of autoimmune diseases (where the immune system gets confused and attacks its own cells). These include: arthritis, allergies, asthma and inflammatory bowel conditions. A low-grade inflammation in the body, in response to an overly primed immune system, can increase the risk of cancer. When combined with an increase in the clotting tendency of the blood, our low-grade inflammation also increases the risk of heart disease.

So changing the ratio of omega-6 to omega-3 in the blood increases the risk of all these modern types of diseases – which were rare before processed food and vegetable oils literally became part of our bodies.

Finally, low-grade inflammation in the body increases TNF-alpha (the inflammatory molecule discussed in chapter 5). We know that TNF-alpha acts to block the action of leptin. Leptin resistance causes a higher weight set-point and – you guessed it: obesity.

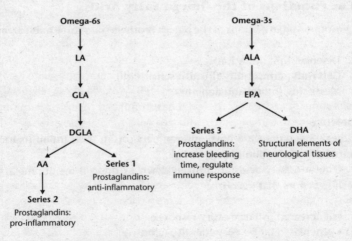

Figure 9.9 The inflammatory chemicals of omega-6 and omega-3

Key: LA – linoleic acid; GLA – gamma-linolenic acid; DGLA – dihomo-gamma-linolenic acid; AA – arachidonic acid; ALA – alpha-linolenic acid; EPA – eicosapentaenoic acid; DHA – docosahexaenoic acid.

Source: Adapted from W. E. Lands (1992). Biochemistry and physiology of n-3 fatty acids. *FASEB J*. 6 (8), May, 2530–36.

Cell Wall Permeability

The flexible, dynamic and fast-moving tail of the omega-3 fatty acid increases the flexibility of the cell wall. This makes the cell wall more fluid and adaptable. Transmission of elements, such as calcium, through the cell wall is faster when there is more omega-3 present. Metabolic adaptation and adaptability are increased. The wall is also more sensitive to hormonal messages from the outside. The opposite occurs when there is a high proportion of the stiffer omega-6 in the cell wall. Adaptability and permeability are decreased. Metabolism is blunted. The cell membrane is less sensitive to hormonal messaging.

An important change in cellular walls with high omega-6 to omega-3 ratios is a *decrease* in the sensitivity of the cell wall to insulin in muscles and leptin in the brain. Higher insulin levels and leptin resistance raise the weight set-point, increasing the risk of obesity.[7]

Messaging (Mood and Appetite)

The omega-6 fatty acids act as precursors to *endocannabinoids*, which are signalling molecules that stimulate the cannabinoid receptors located in the brain. Yes, you may have guessed, these are the same receptors that are triggered when cannabis, or weed, is smoked! The effect of stimulating the cannabinoid receptor when you smoke weed is an elevated, happy mood. If the dose is high enough, you are likely to experience feelings of euphoria as well. We also know what happens about an hour after the weed has stimulated those cannabinoid receptors: a sudden appetite combined with food-seeking behaviour. When food is eventually eaten, it gives a more pleasurable feeling and also enhances any sweet taste.

What happens to our endocannabinoid system when our omega-6 to omega-3 ratios are massively elevated by exposure to the modern Western diet? The excess of omega-6 produces an excess of endocannabinoid messengers and the system is chronically over-stimulated.[8] This is not to say that everyone with a high omega-6 to omega-3 ratio will walk around cheerfully looking stoned – but that the same system, the endocannabinoid system, is stimulated to a low level over long periods of time.

Let's look at the proven effects of high omega-6 to omega-3 ratios on the function of the endocannabinoid system and therefore on our behaviour and health.[9]

- Activation of CB1 (cannabis) receptors leads to increased appetite and calorie intake.*
- The endocannabinoid system is involved in energy balance and over-stimulation of this system leads to obesity.[10]
- Activation of the system enhances sweet taste and also an increase in the release of pleasure and reward (dopamine) chemicals in the brain. Food tastes better and gives you more pleasure.

The message here is that the omega-6 brothers, those stiff and unfriendly wall guards, have a secret cannabis affinity. The more omega-6 in the cell wall, the more the appetite and weight-regulating system will be ratcheted towards weight gain. And gaining weight will be a pleasurable experience because the omega-6 stimulates a more pleasant taste and more rewarding food feelings. That familiar KFC family bucket craving doesn't come from nowhere: it comes from the previous experience of KFC, whose omega-6 is still clogging up your cells and turning out cannabinoids.

Another important health implication of altering our omega fatty-acid ratios is in the effect on their function within the brain. This book does not have the scope to go into detail about this but we should bear in mind that:

- High levels of omega-3 (25 per cent) are normally present within the brain cell membrane
- Changes in our dietary omega-6 to omega-3 intake alter the ratio within the brain
- Severe omega-3 deficiency can cause numbness, weakness and blurred vision
- Low omega-3 levels in the brain are found in multiple sclerosis, macular degeneration and Huntington's disease[11]

---

* Of note is that one of the most exciting and effective drug treatments for obesity appeared ten years ago. This was a drug called Rimonobant and it blocked the CB1 (cannabis) receptors in the brain – the same receptors stimulated by the endocannabinoids originating from omega-6s. Many of my patients taking the drug reported great weight loss. However, Rimonobant was taken off the market after only one year with reports that it could lead to psychosis and even suicide.

- An increase in the omega-6 to omega-3 ratio has been implicated in: Alzheimer's disease, dementia, anxiety mood disorders, suicide.

All these disorders are increasingly common in the Western world.

John Stein, Emeritus Professor of Physiology at the University of Oxford, commented on the increasing ratio of omega-6 to omega-3 fatty acids, saying: 'The human brain is changing in a way that is as serious as climate change threatens to be.'

Omega-6 Blocks Omega-3

Finally, as if the news was not bad enough already, high levels of omega-6 can prevent the body from converting the omega-3 that we get from plants to the more active omega-3 that we get from fish and animals. In other words, if the diet is already very high in omega-6 it doesn't matter how many green vegetables you eat – their conversion into useful omega-3 will be blocked. So we have a compelling argument that, yes, a deficiency of an essential dietary nutrient – omega-3 – is a cause of obesity. This fits in with:

- Epidemiological evidence – populations with a low omega-6 to omega-3 ratio do not suffer with obesity (e.g. the Japanese and any non-Western rural communities), whereas those with a high omega-6 to omega-3 ratio always have high rates of obesity.
- Lipid (fat) cell membrane research – multiple effects of high omega-6 to omega-3 ratios on metabolism have been found, leading to a raised weight set-point.
- What patients say – the omega-6 to omega-3 ratio takes months to change, and dieting will not have an impact on this. This explains weight regain after dieting. When patients move to another country, their weight set-point will readjust to the new food environment. (We will discuss how exactly you can do this, without moving country, in Part Three.)

Our New Deficiency Disease

Just as a lack of vitamin C in the diet *causes* scurvy and leads to *symptoms* of extreme tiredness, personality changes and cravings for food, so the altered omega fatty-acid ratio in the Western diet *causes* a

Calories consumed
Obesity rate
Percentage
40
30
20
10
0
1970
1980
1990
2000
2010
Despite FEWER calories consumed from 2000 to 2010 obesity rates continued to rise. Intake of vegetable oils continued to rise sharply in this period
Lb per capita per year
40.0
30.0
20.0
10.0
0.0
1909
1929
1949
1969
1989
2009
Butter
Lard
Shortening
Oils

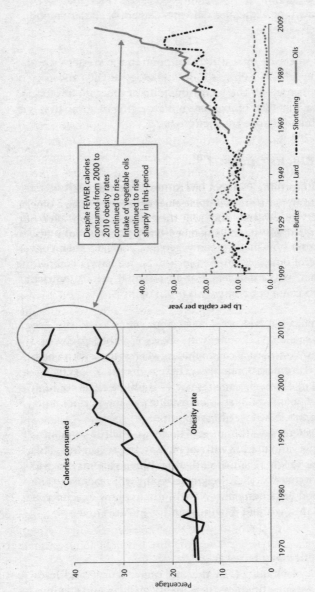

Figure 9.10 Link between the intake of vegetable oils and obesity rates in the USA, 1970–2010

Sources: For calories consumed, see USDA Economic Research Service, loss-adjusted food disappearance; for obesity rates, see C. L. Ogden and M. D. Carroll (2008). Prevalence of Overweight, Obesity, and Extreme Obesity among Adults: United States, Trends 1960–1962 through 2007–2008. *National Health and Nutrition Examination Survey (NHANES)*, June. National Center for Health Statistics; for added fats, see data from the USDA Economic Research Service.

raised weight set-point. Hunger and tiredness are then the *symptoms* of the condition and result in weight gain and, ultimately, the disease of obesity.

Winter is Coming – Evolutionary Adaptation

Our weight set-point is a programmed response to information from our genes and epigenes, and to our past and present environments. This data is used by our bodies to select an appropriate size of 'fuel tank' (or stored-energy reserve) that will help us survive in a future environmental disaster such as famine. When we think of it this way, then the omega fatty acids are acting as messengers, like a proxy from nature, informing us about our future environment.

Examples from nature may provide clues as to how our own omega messaging service evolved, and *why* these messages can alter our weight and immune system. Remember, omega-3 comes from leaves and omega-6 comes from seeds and nuts. In temperate climates, away from the tropics, the levels of omega-3 and omega-6 in the food available will vary as the seasons change. In the spring, as plants start to produce shoots, omega-3 becomes predominant in the diet. In the autumn, when the leaves start to fall, the seeds and nuts are more plentiful – therefore omega-6 will be in the ascendancy. Over a period of weeks and months the cell walls in an animal change in synchrony with their food environment. The omega-6 to omega-3 ratio is lower in the spring and summer and then increases in the autumn and winter.

What effect do these seasonal cellular changes have on some animals? Their behaviour and biology change to prepare them for the approaching winter. As the strength of the sun starts to dissipate in the autumn, so the amount of food energy in the environment will also decrease. Animals have to start adapting to a decreasing supply of food energy in a cold winter environment that requires more heat energy to survive. Their metabolic balance becomes stressed as the environmental signals are predicting less energy in and more energy out. Birds have a logical way of dealing with this: if the energy equation doesn't suit them, they fly south to countries where there is more abundant food energy from the sun. But what about animals stuck on land and unable to migrate long distances?

We know the brown bear will develop a voracious appetite as winter approaches and will gain an extra 30 per cent of its body weight – all as fat – to be burned slowly during the long months of hibernation.[12] It is hypothesized that the levels of omega-6 and omega-3 in the diet, and therefore in hibernating animals' cell membranes, act as a trigger for hibernation, or torpor[13] – more omega-6 leads to a rise in their weight set-point as they prepare for winter months without food. Whilst food energy is still available in the late summer and autumn, the brown bear will gorge until its 'fuel tank' is as full as the weight set-point has calculated is best for it to survive winter. When the cold sets in, the brown bear has another strategy for survival – it will reduce its temperature, and therefore its metabolic rate, by hibernating. Throughout the cold winter months, the brown bear will slowly burn through its energy reserves before temperature signals wake it in the spring.

Another example of behaviour changing in response to environmental signals is that of the chipmunk.[14] As autumn approaches, food availability changes (fewer berries and more nuts). The change in the omega ratio of the chipmunk's cell walls is one of the signals thought to stimulate appetite and weight gain. In addition, it exhibits hoarding behaviour (you may recognize this), bringing nuts to store in its burrow to help it through the winter. The omega-6 to omega-3 ratio also signals when the animal should go into a torpor (a state of very low energy expenditure and metabolic rate, but the animal remains conscious).

The yellow-bellied marmot, a type of ground squirrel that inhabits chilly Canada, spends eight months per year in hibernation. It eats mostly leafy plant foods, grasses, nuts, eggs and insects. If its omega-3 levels are artificially raised in the lab, then it just doesn't get the signal to hibernate.[15] This suggests that, in the wild, a change in the omega-3 to omega-6 ratio triggers the start of hibernation.

Winter Food and Animal Biology

These are examples of hibernating animals whose omega profiles have been studied. We know that there are other environmental triggers that start a hibernation response in wild animals, including changes in temperature, ambient light and vitamin D, but I think that there is a compelling case that changes in seasonal foods also lead to alterations in the behaviour and biology of animals, including humans. According to this theory, the changes we see in response to high omega-6 to omega-3 ratios are in fact primordial protective responses against a future harsh environment. It seems to make sense. As winter approaches, the omegas signal that we need a more accentuated appetite; that we need to seek food more urgently and to savour and enjoy that food more. We need less leaky and more metabolically stable cell membranes to prevent metabolic wastage. We develop stronger immune systems to help fight infection and heal tissues during the winter. Some scientists have speculated that the insulin resistance seen with an increased omega-6 to omega-3 ratio, which causes a higher blood glucose level, is a survival trait inherited from ancient organisms protecting themselves from freezing (the freezing point of water decreases as sugar is added to it). Certainly, some extreme hibernators, such as the yellow-tailed frog (which can actually cause some parts of its body to freeze) still use this strategy.[16]

All these biological responses, which may be part of our ancient evolutionary baggage, if present, were designed to help our survival. In that case, then, a major cause of obesity is not a lack of willpower, or laziness, but the appropriately protective weight-gain response to a change in the environment. Unfortunately, the current changes in our food environment are extreme, not because of the seasons but because of our Western diet. No natural autumn season would produce such a big swing in the fatty acids towards omega-6 and away from omega-3, and no natural autumn would last indefinitely.

This theory, although unproven, does make sense – it fits in with all that we have seen and heard about obesity. It explains all our previous mistakes and clarifies why some treatments work only partially and some don't work at all. It fits in with our weight set-point theory and explains exactly why some of my patients are continually unsuccessful in trying to reduce their weight. But it also explains why – very occasionally – some have been successful. We can understand why some people's weight set-point drifts up or down depending on which country they move to and the omega ratio of the food in their new environment.

But this theory does not explain all cases of obesity. There are other factors in the environment that will alter our weight set-point upwards. These include our habit of snacking and the GI (sugar) index of our food, both of which will cause chronic insulin elevation and obesity. We will discuss this in the next chapter. We also learned, in chapter 3, that recurrent low-calorie dieting will also elevate our weight set-point as a protector against future food shortages. There is also the subgroup of patients, described in chapter 5, whose obesity has become extreme and uncontrollable: these people have developed leptin resistance, which will drive further weight gain even in the presence of positive environmental and dietary changes. We will bring together all these factors and discuss how we can optimize our weight set-point in Part Three of this book.

As scientists and doctors, we sometimes fail to learn from our past mistakes. History should have taught us that in fifty, or a hundred or two hundred years from now our colleagues of the future will be amused by our current convictions and our misunderstanding of the biggest health crisis of a generation. Just as we saw with beriberi and scurvy – the answer is all around us. We just need to see it.

Summary

In this chapter we learned why the *type* of fat that we eat is so important to our health and our weight. We discovered that there are two particular fats – omega-3 and omega-6 – that cannot be made within our bodies and therefore, just like vitamins, are essential in our diet (that's why they are called essential fatty acids). These two fats compete with each other for space on the walls of every cell in our bodies. The amount of each of the two omega fats that we eat reflects the amount seen on our cell walls. And the amount (or ratio) of the two fats on our cell walls has profound effects on our metabolism, our weight and the degree of inflammation within our bodies.

From the 1980s onwards, scientists (and governments) recommended that we change our fat consumption away from natural saturated fats and towards polyunsaturated vegetable oils. Vegetable oils have extremely high levels of the omega-6 fat, which makes them stable and less prone to oxygenation (or 'going off'), and therefore suitable for addition to foods that require a long shelf life. The consumption of seed oils (sunflower, rapeseed, soya) has tripled within three decades. The resulting excessive amounts of omega-6 fats consumed in the Western diet translate directly into the make-up of the cell wall fats. The ratio of omega-6 to omega-3 fats in a population's food supply is also mirrored in their cell walls. The change in the type and amount of fats consumed in the Western diet has resulted in an increase in the omega-6 to omega-3 ratio from a natural level of four omega-6s for every one omega-3 to the current level up to *fifty* omega-6s for every one omega-3.

Increased cellular omega-6 levels cause an increase in inflammation (contributing to a range of Western diseases). Increased inflammation (via TNF-alpha) leads to poorer functioning of insulin and a dulling of the effect of leptin (the hormone produced by fat cells that keeps us thin). Poorer insulin function means that more insulin is required within the blood. Higher levels of insulin also result in a dulling of the leptin signal. All these effects result in a rise in the weight set-point and then, inevitably . . . weight gain.

There is some evidence that the foods that hibernating animals eat before winter sets in act as a trigger for rapid weight gain. The signal comes from the shift of food availability in the autumn/winter towards nuts and grains (omega-6) and away from shoots and leaves (omega-3) – affecting the animal's cellular omega-6 to omega-3 ratio and triggering weight gain. Towards the end of this chapter we speculated whether humans have a similar evolutionary response to 'autumn' foods. In fact, the Western diet has a much stronger effect on our own omega-6 to omega-3 ratio than can be seen in a switch from spring to autumn foods. In addition, the Western diet remains much the same whatever the season. This could produce a permanent signal to gain weight – and be a strong contributor to the development of obesity in some people.

Just like the deficiency diseases of the past – beriberi and scurvy – could a relative shortage of omega-3, compared to omega-6, be an important trigger for today's health epidemic – obesity?

TEN

The Sugar Roller Coaster

Glucose, Insulin and Our Weight Set-Point

Have you ever wondered whether a particular sportsman uses drugs? Maybe they keep on getting better and stronger, despite getting older: they seem to be ahead of the pack and always look great – totally healthy, and their muscles just seem to defy fatigue. What could be their secret? Is it just good genes, eating well and training hard? Or could it be that they are very clever and take some kind of drug which is almost impossible to detect in the drug tests athletes have to undergo?

Many bodybuilders, up to 10 per cent according to a recent report, now take such a drug.[1] It is used to promote uptake of glucose from the blood into the muscles. This means that the muscles can store more energy and work for longer. It is also thought to protect against muscle breakdown. If you need more energy stored up in your muscles than your rivals, then this is the drug. Just one problem though: if you misuse it, it will kill you, within minutes.

Only one major athlete has ever admitted to using this drug. But the testers never caught her – the drug we are talking about is eliminated from the body within minutes, leaving no trace. The athlete, Marion Jones, was the fastest woman in the world and a poster girl for American athletics. She admitted to using *insulin* – along with other drugs – to enhance her performance.

Insulin causes blood sugar to be stored in cells to be used later. Normally it is produced by the pancreas gland in response to a high level of glucose in the blood.* However, athletes can force the issue by taking in extra insulin *plus* extra sugar over a couple of hours (in

---

* Insulin is also produced after eating protein. However, at the same time as insulin is released, protein also stimulates a hormone called glucagon. Glucagon has the opposite action to insulin, so in effect protein remains insulin-neutral.

a process called the hyper-insulinaemic clamp). This overfills their muscles with glucose – and gives them an endurance and performance edge over athletes who don't use the technique. The downside is that if they do not understand that it is essential to take sugar when they take the insulin – if they don't read the small print – then the insulin will consume all the available blood glucose, it will disappear into the cells, and none will be left over to feed the athlete's brain. They can fall rapidly into a coma and die. Despite this risk, I suspect that there are high-profile athletes using this drug, under supervision, in the knowledge that they will never get caught and it will give them an advantage over their rivals.

Insulin is a great drug for getting glucose into the *muscle cells* if you are an athlete. But if you are not constantly exercising, then it has a different effect. Instead of your muscles looking full and ripped, the insulin will force the glucose into your *fat cells* – and after a while your belly, and not your muscles, will look full, and possibly more ripe than ripped.

I saw a patient recently who had been fighting diabetes and obesity for years. He was only twenty-five, but had suffered with diabetes since the age of ten. When he was put on to insulin therapy to treat the diabetes, he noticed that he put on a lot of weight. He hovered between 100kg (15 stone 10lb) *on insulin*, with good diabetic control, to 80kg (12 stone 8lb) *off insulin*, with poor diabetic control. When he felt that his obesity was getting too much, he would stop treating his diabetes, stop taking insulin, and his weight would fall. However, his weight-loss technique was not doing his body any good: he had already started to develop retinopathy (damage to the retina), a complication of poorly controlled diabetes that can eventually lead to blindness.

Weight gain is a well-known side effect of insulin therapy for diabetes. Insulin forces the blood to surrender its energy reserves into fat cells. When insulin is in our bloodstream, the energy gate is only open one way – *into* fat cells, which will never let any energy out. The fat is locked in.

If insulin levels in our bloodstream are high, then we can expect that our weight set-point will be raised. This is what happens in diabetic patients. If insulin is withdrawn – as in my patient who was fighting obesity – weight loss follows. There are many scientific studies that confirm that altering insulin levels will lead to changes in body weight. Increase insulin and you will increase weight; decrease insulin and you will lose weight. Insulin changes the set-point – up or down – and then the weight will follow.

High insulin = higher weight set-point
Lower insulin = lower weight set-point

An interesting study from San Diego, California, confirmed that insulin works on the set-point.[2] The weight of fourteen diabetics was measured as their insulin therapy was slowly increased over a six-month period, until their blood glucose had been controlled. It confirmed that the subjects gained over 8kg (1 stone 4lb). However, when they analysed just how much the subjects were eating while on insulin therapy, they were very surprised: despite the weight gain the subjects seemed to be eating 300kcal less per day than they had prior to insulin treatment. As we learned in chapter 3, insulin causes the master controller of our weight – leptin, the hormone produced from our fat – to malfunction. This leads to leptin resistance and a high weight set-point. The metabolic effects of leptin resistance (in this case caused by high insulin levels) are the same as if leptin levels had been reduced by weight loss from an illness or a famine (or a diet): low metabolic rate. Despite the subjects eating less, their metabolism had slowed down, resulting in weight gain – a perfect example of how insulin drives the set-point upwards and the metabolism downwards. So, as well as acting directly on cells to encourage energy storage and weight gain, insulin also acts indirectly to promote weight gain – by causing leptin resistance.

Insulin → leptin resistance → lower metabolism →
higher weight set-point

What happens when we give a patient a drug to lower insulin levels? Will this have a beneficial effect on their weight? Robert Lustig's research group from Tennessee looked at the effect of reducing insulin levels in obese volunteers.[3] They were given a series of injections of octreotide that reduced insulin secretion from the pancreas. After the course of treatment the group had lost weight (on average 3.5kg or 7.7lb). In addition, their insulin sensitivity (how effective the insulin was) improved. The group reported that their appetite had been reduced by the treatment.

Decreasing insulin level → decreased weight

The Omega Foods That Influence Insulin Levels

We saw in chapter 9 that certain factors influence whether insulin works well or poorly. One of these is the omega-3 to omega-6 ratio. If there is too much unfriendly omega-6 in the cell membrane (because of excessive vegetable oil and grains in the diet), then insulin won't signal properly. You will need to produce more of it to get the same effect. Omega-6 also triggers inflammation and the production of TNF-alpha. We saw in chapter 5 that this also acts independently to reduce the effectiveness of insulin on the cell membrane (as well as causing leptin resistance). So, the fatty acids in the Western diets – the ones used in oils and shortenings, including the supposed heart-healthy vegetable oils – cause us to need more insulin and therefore our weight set-point rises. High omega-6 fats in the diet, and the TNF-alpha inflammation that they cause, act indirectly to make the cells less receptive to insulin (they cannot sense it as well) – meaning that more insulin is required.

Western diet → high omega-6 to omega-3 ratio → insulin resistance → more insulin → higher weight set-point

High omega-6 to omega-3 ratio → inflammation → insulin resistance → higher weight set-point

A Teaspoon of Sugar . . .

Our brain needs sugar – it needs the glucose in our bloodstream to function.* We need to maintain an optimum blood level of this precious brain fuel: too little and we will fall into a coma, too much and the glucose causes inflammatory havoc. But the actual amount of sugar carried in our blood is surprisingly small.

Most people have 5 litres of blood pumping around their bodies. How much sugar does it contain? Imagine filling a bucket with 5 litres of water. How much sugar should we add to our bucket of water to make it the same sweetness as our blood? The answer may amaze you – we need to mix in just one teaspoonful of sugar to reach our optimal blood level of 80mg/dl. We have large reserves of sugar stored in our muscles and our liver, but only one teaspoonful is present in

---

* Unless we are in starvation mode, in which case our cells break down fat and produce a glucose substitute in the blood called ketone bodies.

the whole 5 litres of our blood. Our glucose-transporting hormones, particularly insulin, are essential in maintaining this level of glucose in our blood to keep us alive and healthy.

Glucose is the End-Product of All Carbohydrates

After we eat any food that contains carbohydrates, glucose starts to be dumped into the bloodstream. The pancreas senses this and starts to secrete insulin. Insulin's job is to direct the blood glucose into our cells (mostly our fat cells). When insulin levels are high, the body switches to storage mode: insulin forces glucose from the blood and into fat cells, where it is converted to triglycerides.* Once the blood glucose settles back down to our single teaspoonful level again, the insulin disappears – it is no longer needed.

The amount of insulin that the pancreas secretes is proportional to the amount of glucose being dumped into the circulation. And the amount of glucose being dumped into our blood, and the speed at which it is being dumped, depend on what we have eaten. Therefore insulin levels are directly influenced by the *type* of food that we eat. Food or drink with a high sugar content (think Coca-Cola) will produce a surge in insulin. Foods containing complex carbohydrates (think of a stick of celery) that need time for the bowel to break them down will infuse their glucose much more slowly into the system and therefore produce a longer but less intense secretion of insulin – a trickle.

As an example, a teaspoonful of sugar contains 4.2g of sugar – the same amount as in one eighth of a can of Coke, and the same amount contained in four sticks of celery. If we drank a mouthful of Coke, this would infuse sugar into our blood within minutes, causing a surge in our blood glucose (doubling the amount); therefore a large spike of insulin would be needed to deal with this. If we ate four sticks of celery, our intestines would need up to an hour, or two, to break down the complicated chains of carbohydrate contained within the celery into simple sugars. Therefore the blood-sugar level would rise extremely slowly. Insulin would also rise slowly to deal with the sugar. The total amount of insulin produced to deal with the Coke sugar would be exactly the same as the total amount of insulin required to

---

* High triglyceride levels are implicated in the risk of heart disease. High sugar intake causes the production of triglycerides. It is increasingly accepted now that sugar in the diet, and not natural saturated fat, is the true heart disease risk (see Appendix 1).

deal with the celery. Remember, total insulin secretion over a long period of time is used as part of our set-point calculation: the more insulin, the higher the weight set-point. With carbohydrates, it is not the type of food that matters when calculating insulin and the set-point, it is the total amount of sugar.

Highs and Lows on the Roller Coaster

To expand on this a little, sugar is sugar whether it is in Coca-Cola or celery. However, if the food (or drink) that is taken in lets go of its sugar too fast, and if that food has a tremendous amount of sugar in it, then it will cause a violent spike in insulin secretion. The large surge in glucose entering the bloodstream is followed by a large surge in insulin. The insulin works to open up the fat cells to suck the sugar into them but . . . here is the catch: because of the insulin surge, *too much* glucose leaves the bloodstream, leading to the blood-sugar level *falling*. This sets off alarm bells in the brain which is dependent on glucose to function. The alarm gives us a feeling of anxiety and a very strong craving for any food containing sugar. The brain is stepping into critical survival mode – 'Get sugar and get sugar fast!' is the message. We cannot ignore these warning messages; we have to act on them, and therefore we seek sugar wherever it is. Fortunately, in the Western environment, sugar surrounds us. Unfortunately, sugar drenches most of our foodstuffs. What the brain wants is less than a spoonful of sugar, maybe half a spoonful; what it gets is more likely to be a low-fat blueberry muffin (containing nine spoonfuls of sugar). Sugar floods the bloodstream again and insulin is called into action once again – massive surges of it – and the cycle is repeated.

This is the classic sugar roller coaster. Eat a highly refined sugary breakfast (cereal and/or toast + orange juice) and you will get a sugar spike soon afterwards. By mid-morning your blood sugar levels will have plummeted due to the large reactive spike in insulin forcing sugar away from your blood and into your fat cells. It is at this time – in the mid-morning – that you get that terrible craving for more sugar. The coffee shop is calling you – *blueberry muffin time*! Wow, up again, great feeling. Then, pre-lunch: scary, scary, the adrenaline is dropping; low sugar again. Let's hit the supermarket and get a meal deal: sandwich, crisps (you need those endocannabinoids), a Coke. Yes! Up and down, up and down – all day . . .

This is the thrill of the sugar roller coaster. But this is a new ride.

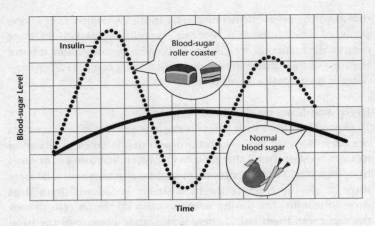

Figure 10.1 The blood-sugar roller coaster

Source: J. Brand-Miller et al. (2009). Glycaemic index, postprandial glycaemia, and the shape of the curve in healthy subjects: analysis of a database of more than 1,000 foods. *Am J Clin Nutr*, 89 (1), January, 97–105.

Before the US government's dietary guidelines in 1977 (the McGovern Report), many people would have consumed a hearty breakfast of fried eggs and bacon or sausage – the full English. This would usually carry people through the morning until a small lunch was contemplated – there was no thrilling high-sugar rush and no subsequently scary low-sugar dive. But we were then told that the saturated fat in the full English breakfast was going to kill us, so we had to stop eating it. Instead we ate low-fat (high refined carbohydrate) breakfasts and took our early-morning seats on the sugar roller coaster.

'The Snack You Can Eat Between Meals Without Ruining Your Appetite'

The food industry had a new opportunity after the publication of the 1977 dietary guidelines. People were consuming new 'healthy', low-fat, high-sugar breakfasts and did not feel so well mid-morning. They craved a mid-morning snack, but it was not normal eating behaviour to snack between meals. Our benevolent food industry came to the rescue.

Sensing the new need for people to top up their sugar mid-morning and mid-afternoon, and also sensing a money-making opportunity, the food industry set about changing our eating habits. It had already started to change the food we ate – but this new type of food itself required that we change our eating habits as well. Eating three square meals per day, and embracing a normal hunger (working up a healthy appetite) between meals, would need to change to accommodate the new type of high-sugar food we were consuming.

Slowly the advertisers started to suggest that it would be a good idea to treat children to a snack between meals, that it would be good for their concentration at school, that it would not ruin their appetite. New light and fluffy-looking chocolate snacks were developed, and we were assured that this was 'a snack that you can eat between meals without ruining your appetite'. Slowly, over the years, our eating behaviour changed inexorably until it was not at all unusual to see people enjoying a snack between meals – something that would have been seen as odd prior to the 1970s. Eventually snacking became normal – even snacking in public was considered normal behaviour. Our behaviour had been changed, and a multi-billion-pound snack industry had been spawned.

The cumulative effect of the sugar roller coaster is that we consume much more glucose, much more sugar, than we would normally crave from a healthy balanced diet. By eating so much highly processed food, and with the new culture of satisfying ourselves during the day on snacks, we are increasing the total amount of insulin that is required to cope with these waves of glucose. As we know, the total amount of insulin that we use is reflected in our weight set-point. Higher than average insulin levels over several weeks will increase your set-point, and weight gain will follow. Whereas lower than average insulin levels over several weeks will lower your set-point and effortless weight loss will follow.

So it is not about the total *number of calories* that we are consuming with our constant snacking – if these calories counted as part of an energy in/out equation we would quickly balloon to 200kg, and remember that the massive increase in the number of calories that we are consuming in our sugary snacks is burned off by ratcheting up our metabolic afterburners. This is our normal metabolic adaptation to over-eating (as discussed in chapters 1 and 3). The crucial point – which contributes to people gaining weight – is the *effect of insulin on the weight set-point.* Insulin dulls the leptin signal and the body gets the message to store extra energy. As the insulin level increases, so leptin resistance appears. If our master weight controller, leptin, is not heard, our set-point will rise and our weight will follow it.

Once we decrease our average insulin level, our weight set-point, and then our weight, will fall. In Part Three we will look at ways that we can do this by adjusting the way we eat. We will also learn, with the help of my expert eating psychologist, ways to wean ourselves off that addiction we may have to sugar.

How Does Alcohol Affect Our Weight Set-Point?

Alcohol is made from sugar. Those of us who drink regularly are reminded by the media just how many calories are contained in our evening glass of wine. We are told that alcohol contains 7kcal per gram – much more than the energy in a gram of carbohydrate or protein (4kcal), and almost as much as a gram of fat (9kcal). A large glass of wine or a pint of beer contains over 200kcal, the same as a large slice of pizza. Have a couple of glasses and you are reaching 20 per cent of your recommended daily calorie intake. If you are celebrating in a nice restaurant and have an aperitif, followed by a couple of glasses of wine, then the calorie content from your alcohol (600kcal) can be more than from the food.

We know that alcohol can cause many serious illnesses, from cirrhosis of the liver to heart disease and cancer, but how much does our alcohol consumption contribute to our obesity crisis? When we count the calories, it looks bleak. For adults, the average weekly alcoholic calorific intake in the UK is over 1,800kcal – enough to make you think about quitting. But this book is all about *not* counting calories and instead questioning how different factors affect us metabolically. So, for now, let's push away those guilty thoughts about how many calories we consume in alcohol and think about how it influences our weight set-point.

The Drinking Man's Diet
In the 1960s an interesting diet briefly became popular. *The Drinking Man's Diet*, a book by Robert Cameron, self-published in 1962, suggested that if you substituted the calories consumed from sugar and starches and replaced them with alcohol you would shed some excess weight. The diet was based on the observation that many alcoholics fail to gain weight despite consuming massive amounts of calories from alcohol. How do heavy drinkers metabolize or burn off these excess calories? The question has baffled nutritionists for years.

In 1991, researchers from Mount Sinai in New York answered the puzzle.[4] They studied a group of alcoholic men whose weight was stable on 2,500 calories of food daily. When they added 2,000 calories of alcohol to the 2,500 calories of food, they found no weight gain. The alcoholics were somehow able to burn off the excess calories contained in the alcohol and maintain their weight. When the researchers replaced the alcohol with 2,000 calories of chocolate, the subjects gained weight. They found that in heavy drinkers a cellular mechanism occurring in the liver seemed to be able to literally burn alcohol calories, creating increased thermal energy (in a similar way to the thermogenesis described in chapter 3).* Subsequent research showed that as well as turning alcohol calories into heat, heavy drinkers could also increase their metabolic rate by stimulating their sympathetic nervous system (just as in our over-eating experiments in chapter 1), resulting in a faster heart rate and higher blood pressure – all helping to dissipate energy.[5]

Most doctors will notice that when they examine a patient who has recently been bingeing on large amounts of alcohol their skin feels hot. No fever, or temperature – just hot. These studies explain this phenomenon. In heavy drinkers, alcohol is broken down and converted, not to chemical energy to be used or stored in the body, but to heat energy which is radiated through the skin. This is the reason a heavy drinker, during an alcohol binge, will not feel the cold, even in sub-zero conditions.

If, as it appears, much of the calorie energy in alcohol is burned off as heat, and not stored in alcoholics, what about more moderate drinkers or occasional drinkers? How is alcohol energy processed in most of us? Alcohol energy itself cannot be stored in the body in the same way that fat or carbohydrates can be. The body treats it like a poison, breaking it down first into a chemical called acetaldehyde (which causes your hangover) and then into acetate (vinegar's essential component), before finally being converted to carbon dioxide and water. During the breaking down of alcohol, small particles of nicotinamide adenine dinucleotide (NADH) are released. These particles contain the energy to charge up those ATP micro-batteries (discussed in chapter 7) in our liver cells. So some energy is produced during alcohol breakdown, but none is directly stored and must be used immediately. The liver cells suddenly have lots of

---

* This process that takes place in the liver is known as the microsomal ethanol oxidizing system.

cellular ATP batteries charged, lots of energy on board. During the period that alcohol is being degraded (averaging 2 units per hour), the liver uses this free supply of energy. This means that it does not have to use its normal source of energy: fat. Therefore fat starts to accumulate in the liver cells, resulting in a condition called fatty liver.

So we now know a side effect of metabolizing alcohol can be the risk of developing a fatty liver, but what about our weight and our waistline? Large studies on the effect of alcohol on weight seem inconclusive.[6] Some show that it can cause weight gain, others that it has no effect on body weight at all. One study showed weight loss in women if they began to drink alcohol.[7]

If alcohol causes significant weight gain, then we would expect that the countries with the highest per capita consumption of alcohol would also be prominent in our obesity league tables. The Baltic nations, Russia and Eastern European countries make up the top 10 in our alcohol league table; France and South Korea appear further down. Whereas the obesity league tables are dominated by the Pacific Islands, and the *non-drinking* Middle Eastern States – Kuwait, UAE, Qatar, Bahrain and Saudi Arabia. None of the nations in the top 20 alcohol league table appears at the top of the obesity league tables, and the obesity league table is dominated by non-drinking countries. On that basis, there does not seem to be a compelling link between alcohol and weight.

Beer Belly

This finding doesn't fit in with our common, day-to-day observation that people (particularly men) who spend a lot of their lives in the pub develop a 'beer belly' – an excess accumulation of fat around the waist. One explanation for this could be the effect that alcohol has on the steroid hormone cortisol. We now know that alcohol causes higher levels of cortisol to be produced.[8] Cortisol is normally produced in response to chronic stress. Patients who have a condition called Cushing's syndrome, where excess cortisol is constantly produced, or patients who take long-term steroid tablets for arthritis or other inflammatory diseases, will develop changes in the distribution of fat around their body. These include increased abdominal fat (described in textbooks on Cushing's as a 'pot belly'), a round face (moon face) and skinnier arms and legs. As regular heavy drinkers increase their cortisol levels, they too will eventually develop a similar appearance to patients with Cushing's syndrome – a large 'beer belly'.

Alcohol, Insulin and Appetite

Interestingly, alcohol can *improve* insulin's function, making it more efficient. But the side effect of this is that when we drink alcohol it can lead to lower levels of blood sugar. The brain senses this and tells us to go and eat – producing the late-night craving for an after-drink kebab, or the early-morning need for a fry-up.

Alcohol, because of its effect on blood-sugar levels (and cortisol), can make us *eat more*. The calories in the alcohol itself are used by our bodies very inefficiently, but alcohol heightens our appetite, telling us to eat high-calorie foods. If the extra food that we take in is high in sugar, wheat or vegetable oils (i.e. Western food), then those foods, if consumed regularly, will raise the set-point.

Alcohol and Our Weight

If alcohol leads us to eat more, then why don't the countries with the highest alcohol consumption appear in the obesity league tables? The answer comes from the quality and type of foods available to those populations. The people of the Baltics, Eastern Europe, Russia, France and South Korea have not yet fully embraced highly processed Western foods, so despite their high alcohol consumption, even when they do over-eat it is with foods that do not raise the weight set-point. Therefore weight does not become an issue.

If the calories in alcohol are not stored, then how do we explain the common observation that people lose weight when they stop drinking? They tend to be people who drank quite heavily before quitting. When they stop drinking alcohol, their appetite returns to normal, and their eating behaviour improves. This, combined with a drop in their cortisol levels, leads to a decrease in their weight set-point and subsequent weight loss.

Is Quitting Alcohol Good for Weight Loss?

We now know that the calories in our glass of wine or vodka and tonic are used by the body very inefficiently. Alcohol, despite containing 7kcal/gram, contributes much less energy than carbohydrate's 4kcal/gram – hence the Drinking Man's Diet and its brief popularity. The energy in alcohol is released because our body is trying to degrade this poisonous substance. Even in moderate drinkers some of this energy is lost to thermogenesis and heat dissipation. The rest of the energy produced while alcohol is being broken down can power the liver. This creates a moderate energy saving and the laying down of the fat that would normally have been used.

In moderate or heavy drinkers, any weight gain due to their drinking could be secondary to the heightened appetite that alcohol produces, leading to poor food choices. Once the diet deteriorates and more processed Western foods are eaten, the weight set-point will rise. In addition, these drinkers will have higher cortisol levels, causing fat to be distributed in their body to the abdomen. If these changes have affected your weight, then by cutting down significantly your alcohol intake or by quitting completely, your weight set-point and your weight will fall. If you are an occasional or light drinker, and if you make good food choices when alcohol tells you to eat, then it is unlikely that it is having a significant effect on your weight set-point.

Summary

In this chapter we have looked at the effect that insulin has on our weight. We learned that when this hormone is used as a treatment for diabetes, it automatically causes weight gain. When it is withdrawn, weight loss occurs.

Insulin is released by the pancreas when we have too much sugar (or refined carbohydrates like wheat) in our blood. Our blood-sugar levels then spike, resulting in a powerful insulin response. The high insulin causes the cells to suck too much sugar from the blood, which leads to blood sugar that is too low, and then causes a strong craving for more sugar. This 'sugar roller coaster' – fluctuating blood-sugar levels throughout the day – increases our average daily insulin levels. Just as if insulin was given to you as a treatment, this leads to weight gain.

The US government guidelines to reduce saturated fats in our diets have led many food companies to increase the amount of sugar in their ingredients to keep their foods palatable. The result has been a 20 per cent increase in the amount of sugar consumed by the general population since 1980. In addition, as discussed in the previous chapter, the recommended dietary changes led to vegetable oil consumption skyrocketing and resulted in an increase in the omega-6 content of our cells. The result is a decrease in the efficiency of insulin, meaning even more insulin is now needed.

The recent changes in our diet and our new snacking culture have also led to much higher average insulin levels – producing a higher weight set-point for most of the population. Changes to an individual's insulin profiles can be made by changing dietary preferences. Lowering insulin levels by dietary changes will lead to weight loss. Part Three of this book will provide instructions on how to do this.

ELEVEN

The French Paradox

Saturated Fat, Nutritional Advice and Food Culture

We have been told two things about fat by the nutritional scientists over the last forty years:

1. Fat makes you fat.
2. Saturated fat causes heart disease.

There is now a growing body of evidence that both these pillars of dietary advice are in fact built on shaky foundations (see Appendix 1). As the French, the Maasai and the Inuit have shown us, fat does not make you fat.

Fat has had two perception issues. The first is its energy efficiency. Yes, it does contain more energy calories by weight than other food categories (compared to carbohydrates and protein). It was therefore assumed that people would be consuming more calories if the food was fatty; it was assumed that satiety, the feeling of fullness that stops us eating, was due to the volume of food eaten. However, this is an old-fashioned way of looking at it – unless you happen to be in the middle of the eating challenge show on American TV, *Man v. Food*. In fact, we now know that fat causes a much more immediate and more pronounced satiety response than carbohydrate. When we consume fat, it triggers a strong release of the satiety hormones discussed in chapter 4 (peptide-YY, or PYY, and GLP-1); the hormones then act on our weight-control centre (in the hypothalamus) to stop us eating.[1] The research explains the observation that when rats are fed diets of calorie-dense foods, they do not suddenly start consuming more calories; instead, they control the amount of food consumed based on calories and *not* food volume.[2] The second issue is its unfortunate name – FAT. If a marketing person had coined this name, they would have surely lost their job a long time ago. If foods were to be named for their effect on the body, then sugar should probably be given the name 'fat'. Maybe fat could be called 'strength' or 'vitality'.

Nutritional science, which demonizes saturated fat, is a relatively new speciality. Unfortunately, the advice that the nutritionists have given us has caused more harm than good.* Much of the research is based on poor data (dietary recall questionnaires are notoriously inaccurate) and is sponsored by the food industry (you can imagine the conflict of interest here).

The French Paradox

In Western Europe today, there is a great example of a population whose weight set-point doesn't seem to have gone up as far, or as fast, as its immediate neighbours. The population of France consumes more saturated fat (and more wine) than neighbouring countries, yet it has somehow avoided the worst of both the heart disease epidemic and the obesity crisis. How can this be? The food scientists and nutritionists do not have an answer. They have called it the French Paradox because they cannot explain it. But this may be because their way of thinking about both obesity and saturated fat is flawed.

The French are proud of their national food culture. It is being eroded by the spread of other Western food products, but they are trying to hold out. Despite their access to 'Westernized' food, they mainly eat fresh ingredients. They cook; they do not avoid saturated fat as this is part of their traditional diet, but they do not have the culture of snacking. They are proud of their cuisine. When you have savoured a hearty breakfast, lunch and dinner – with fats that make you feel full, and no empty carbs – you do not feel the need to snack. This is the reason the French remain slimmer than the population in the rest of Europe and the USA – by consuming fresh, unprocessed foods (producing a better omega ratio), fewer

* Nutritionism, a term used by Michael Pollan, an American journalist, in his book *In Defence of Food: An Eater's Manifesto* (2008) to describe the ideology of scientific food reductionism, attempts to understand food by breaking it down into its constituent parts – carbohydrates, fats, vitamins, minerals etc. – and then examines those constituents and tries to work out what is good and what is bad. Nutritionism ignores food culture and suggests that nutritional science can give us the perfect eating advice. Unfortunately, today most of the advice is counter-productive to human health.

carbohydrates and more fat (therefore a better insulin profile) and no snack culture (again improving insulin profile). A few years back, the French government imposed a cap of thirty-five hours for the working week; regulations to give employees the 'right to disconnect' from work emails after a certain time of day have also been rolled out. On top of this, maybe the French prefer an early night with their loved ones or meeting at terraces and bistros for a chat (improving their levels of cortisol and melatonin) to spending hours commuting, then bingeing on Netflix. The end result? A much lower weight set-point and, as a bonus, an improved quality of life. When you eat the correct foods, avoid snacking and enjoy and embrace a lower stress life, there is no need to count calories – your set-point will be within a healthy range and your metabolism will keep it there.

The French ignored the American nutritional advice to cut the cholesterol from their diets. They continued to enjoy cheese and steak and cream while the rest of the Western world switched to vegetable oils, refined wheat and sugar – and got fat. There are other cultures that probably did not heed or hear the Americans' advice: for example, the Maasai tribe in Kenya who just eat meat, blood and milk and remain healthy and very slim despite their extremely high-fat diet. And the Inuit in Greenland, who still consume large quantities of seal and whale meat and blubber, almost pure fat, also avoided the obesity and heart disease crisis.

The Food of the Earth

Food culture plays such a critical part in the health of a population. From the raw fish and rice of Japan to the vegetables and noodles of Indonesia, from the salads, pasta and olive oil of southern Italy to the steak and red wine of the French, the wild meat and tubers of the Tanzanian Hadza to the spicy curries and dal of India – food cultures are so diverse, not only in what people from different countries eat, but how they eat (chopstick, hands, spoons or knife and fork). But all food cultures have two important features in common, aspects that are gradually being lost in the West.

First, food cultures develop over many generations; they are insep-arable from local traditions and, most importantly, they encourage social interaction (through preparing, cooking and eating food together). When the Western world embraced nutritional science, it threw away generations of accumulated food wisdom. Food culture is much more important to health than the individual constituents of the foods that are eaten. It is integral to a happier, more active and more content population – all factors that are ignored by the food reductionists who, by focusing on the benefits/dangers of individual nutrients (i.e. carbohydrates or fats or vitamins), are then able to rubber-stamp refined and fortified highly processed foods.

The second similarity among diverse food cultures is the need to cook fresh and local ingredients. By definition, local food culture means picking, preparing and cooking locally grown food – this means food that is not imported from far away (and therefore not preserved). One of the advantages of freshly grown foods is seasonality. The type of food available depends on the time of year, whether it is spring, summer, autumn or winter. Seasonal foods ensure food variety. Once we have grown bored with summer leaves, the warming winter bean broth will be extra appealing. Seasonal foods are also entwined with our seasonal traditions: Halloween (pumpkin), Christmas (parsnips, Brussels sprouts and swedes), Diwali or Thanksgiving (autumn and winter feasts). Food culture provides a variety of healthy local food throughout the seasons. Fresh local produce, cooked traditionally and with love, and eaten with family and friends – that is true food culture and that is what we have started to ignore with our factory-produced, industry-marketed foods and nutritionalist ideology. Your great-grandmother would not have known what a 'superfood' was, but she would have taught you how to make a steak and kidney pie – just as her mother had taught her.

TWELVE

The Miracle Diet Book

Why You Should Stop Dieting

Bariatric Surgery Outpatients Clinic, London, 2015

'Weight Watchers, Slimming World, the Atkins Diet, the South Beach Diet, LighterLife, the Rosemary Conley Diet, the Dukan Diet, the Red and Green Diet, the Cabbage Soup Diet . . .' My pen moved rapidly across the page of notes, trying to keep up with the list of diets Mrs Thompson had tried. 'And all of the rest, I can't remember them all,' she finished. I looked up. 'Do you want to know why they didn't work for you?'

The New Miracle Diet

A great way to make money is to produce a diet book. All that is needed is a new angle on calorie restriction – something that has not been tried before. Something that people will really think is the solution to their troubles. Add in testimonials of satisfied dieters; if possible mention how the diet changed their lives. A C-list celebratory endorsement comes in handy. The next step is to reproduce a snippet of the new diet book in a tabloid newspaper and then arrange an appearance on daytime TV to explain the miracle discovery.

Here's how a 'miracle diet book' will sell after the initial media splash:

Phase 1: The diet book inspires its readers to go on the new, 'cannot fail', diet. The readers follow the diet, which involves restricting calories. Lo and behold they lose weight, usually between 3 and 7kg (7lb to 1 stone) in the first few weeks! This is the secret: 3–7kg will get the readers' weight loss noticed at work. Their friends in the office will see this and the readers will inform them of the new miracle diet book that has finally saved them.

Phase 2: Ten of the office workers go and buy the book and lose weight. They tell all their friends, relatives and neighbours.

Phase 3: Interest spreads like a dodgy Ponzi scheme, expanding to social media chatter about the book (it could go viral at this stage!) and hundreds more copies are sold.

Phase 4: A few months after the launch is the quiet phase. Book sales may hold up for a little while, but the chatter and interest from the diet is dying down. Everyone, and I mean everyone, has started to regain the weight that they lost. Fortunately, none of the weight gainers are very vocal about this. None of them are annoyed with the book, because they blame *their* poor willpower.

Phase 5: This is the stage at which most readers *weigh more than when they purchased the book*. But this will not stop them looking for the next miracle diet book (they normally appear every six to twelve months). After all, the diet did work – just not for long. It was a short-term miracle diet.

Phase 6: You can imagine the author of the diet book returning to chilly England from their new Caribbean beach-house (bought with the book's proceeds). They notice someone else's new bestselling miracle diet book prominently displayed in the lifestyle section of the local Waterstones . . . and spot their book hidden quite far down the shelves . . . and now sporting a 2-for-1 sticker.

For each of the pounds that were made from each book sold there will be at least as many extra pounds of fat carried around by the readers as they regain the weight they lost – and then gain even more. But it's the perfect crime because the poor readers will think that it is their fault that they put on the extra weight. They will not know that it is a normal metabolic consequence of dieting: the weight set-point is raised in the long term, in the aftermath of weight loss from a low-calorie diet.

The reason there are so many diet books out there is that none of them considers the master controller of our weight: *the weight set-point*. That's why none of the diet books seems to work long term, and that's why there are so many failed miracle diet books cluttering the lifestyle section of your bookshop. Your weight will always return to its predetermined set-point.

When I ask my patients which diet they lost the most weight on,

LighterLife wins the prize in most instances. This ultra-low-calorie food-replacement diet produces outstanding short-term weight loss – however, ALL my patients (without exception) reported to me that within a few months of finishing the diet they had regained 'all their weight and much more'. This simply confirms the conclusion we reached earlier in the book: the more extreme the diet, the more extreme the metabolic and appetite response will be – the weight set-point always wins.

These findings fit in with our weight set-point theory of weight regulation. When food might become scarce at any time, our food-seeking behaviour signals and our enjoyment of high-calorie food are ratcheted upwards and remain there. Diets tend to be short-term – 'I'm on a diet *at the moment*.' Because we are disciples of the simple energy in/out equation we think that we can have a period of negative energy balance for a few weeks, lose weight, then come off the diet when we have reached our ideal weight. This is the underlying premise of all diet and diet and exercise plans – they are short-term, rapid fixes to get the weight off. Because these low-calorie diets can have unpleasant side effects, they cannot be continued in the long term. We have an idea that when we come off the diet we will try and curb some of our previous dietary bad habits – perhaps eating less fast food than before – and maybe we vow to continue with the gym membership. We think that by doing this we will keep the weight off.

A Dieter's Body is Different from the Body of a Non-Dieter

What we don't appreciate is that by the very nature of dieting we have changed our bodies. After we have lost 5, 10 or 20kg in weight by dieting, *biologically* we become a different person. We have adapted to our new low-calorie environment and developed a low metabolism. And remember, our bodies cannot tell the difference between a voluntary diet and a food shortage or famine.

The more weight that we have managed to force off by low-calorie dieting, the slower our metabolism will get, and the more our appetite hormones will be screaming at us not to walk past Starbucks. Our new, metabolically hyper-efficient and hyper-hungry bodies will not *maintain* the weight loss with our original maintenance plan. Cutting down on fast food and going to the gym once or twice a week is not going to be enough any more. In order to maintain the weight loss, we will need to be much more aggressive with our energy in/

out equation: cutting calories further and further, and flogging our poor bodies in the gym even harder. The longer we fight to keep that weight off, the more aggressive the body gets to prevent us – because we are fighting against our weight set-point.

Our weight control is subconscious – just like our breathing. We don't have to remind ourselves to breathe and we shouldn't have to consciously worry about our weight, providing our body is happy with its environment. Just as we can override our subconscious control of breathing in the short term, by holding our breath, so we can also override our weight set-point temporarily by dieting – but when the body senses we are going away from ideal conditions it will intervene to change your behaviour. The longer you hold your breath, the more unpleasant the feeling. Eventually you are screaming inside for release; you know you cannot win. You start to breathe again, and the pain goes away.

A similar protective mechanism comes into play when we diet. As you finally step back into Starbucks and smell the reassuring aroma of coffee – and order the Mocha and glazed doughnut 'snack' (710kcal) – you can breathe easy again: the set-point has won another battle against your conscious will.

Why Not Skip the Diet and Go Straight on to the Maintenance Phase?

Your initial plan was to lose weight on the miracle diet and then, once you had lost the weight, be a bit more sensible with your eating habits and gym attendance. But what would have happened if you had skipped the diet and gone straight on to your lifestyle changes? Maybe you were lazy before and ate fast food twice a week rather than cooking a meal. Maybe you were so tired that you rarely went to the gym.

OK, so now you have decided not to go on a diet, but instead to cook better-quality food and exercise a couple of times a week. Note here that there is no calorie-counting involved in this change. You are just altering a couple of habits for the better. The outcome would be – *no* dramatic weight loss; probably no weight loss for weeks or months (in fact if the exercise you did was weights then you might even gain some weight from your increased muscle). However, if you continued with your better dietary and exercise habits, eventually the message would get through to your weight set-point after some months. Your set-point would go down and maybe many months later you would be 4.5kg (10lb) lighter. A year down the line your weight might settle at 10kg (22lb) lighter. But, unlike the weight loss through dieting, your body would be happy with this weight loss. Your actual weight would be matching your weight set-point, so you would not have a voracious hunger or low metabolism. In fact, because you are fitter, your metabolic rate would increase and as time went by everything will get easier and easier. Your body would be at peace at this weight.

This is an example of weight loss by reducing your weight set-point. Weight loss by changing your habits and environment (e.g. walking to work, doing a sports class or going to the gym a couple of times a week), and therefore changing the signals that you are sending to your weight-control centre, is the only way to successful, sustained weight loss. If you talk to anyone who has lost weight and, more importantly, managed to maintain their weight loss over a long period of time, you will see that they have done this by altering their daily habits and thereby lowering their weight set-point. We discuss practical ways to reduce the weight set-point in Part Three of this book.

New Diet or New Life?

If we compare two people who have lost 10kg in weight – one from low-calorie dieting and one from lowering their weight set-point by lifestyle changes – we can see two different people.

The low-calorie dieter may reach their target weight loss much more quickly than the lifestyle changer, but their weight set-point will not have changed; in fact, it could be slightly higher because of

10kg (1½ stone) weight loss

	By dieting	By set-point
Metabolic rate	Decreased	Increased
Appetite	Increased	Normal
Satiety	Decreased	Normal
Fatigue	Yes	No
Quality of life	Poor	Great
Long-term outcome	Weight regain to higher level than before diet	Sustained weight loss

Table 12.1 Comparison of two people who have lost 10kg – one through dieting; the other by adjusting their weight set-point

the diet. It would be more and more difficult for the dieter to maintain the weight loss over time. In contrast, the lifestyle changer would have a weight set-point matching their weight. The weight loss would have taken considerably longer, but it would be sustainable – in fact, it might get easier over time, as fitness and metabolism increased.

What about the commonly used diets? How might they affect our weight set-point?

Low-Calorie Diets

LighterLife, the Cambridge Weight Plan and the SlimFast diet are all examples of very low-calorie diets (600–1,200kcal/day). These types of diets generally use meal-replacement shakes, soups and low-calorie snack bars (which you have to buy from the company) and therefore by definition are not sustainable in the long term if you want to have a good quality of life (i.e. be able to eat). We have discussed the metabolic changes that occur after low-calorie diets. In the long run they *raise your set-point*, meaning that when you come off the diet you will regain all your lost weight and then some more until your new weight set-point is reached.

Low-Fat Diets

Slimming World's regime is based on a low-fat diet. It does not restrict the amount of food that you can eat but advises you to replace high-fat foods with lower-fat, filling foods. It is based on the flawed assumption that fat makes you fat (see chapter 11) and separates foods into three categories: Free Foods, Healthy Extras and Syns. Free Foods are, as the name implies, free to eat, in unlimited quantities. These include vegetables, lean meat and fruits. But they also include unlimited carbohydrates in the form of pasta and rice. They advise limited amounts of Healthy Extras, including dairy products and also cereals and bread. The Syns are 'treat' foods, like biscuits and chocolate. Between five and fifteen Syns can be taken every day.

The benefits of Slimming World's diet as far as the set-point is concerned are that it encourages home cooking and the consumption of more natural foods. The drawbacks are that people may switch to a much higher carbohydrate content of the food consumed to the detriment of their insulin profile. All in all, dieters who succeed using Slimming World's diet are probably cutting down on ultra-refined carbs (contained in the Syns), and this helps lower the weight set-point slightly.

Low-Carb Diets

The Atkins, Paleo and Dukan diets are based on low-carb eating. Once carbohydrate intake is reduced to under 20 grams per day, a process called ketogenesis occurs.

There are many celebrity advocates of ketogenic dieting, such as LeBron James, Kim Kardashian and Halle Berry. I think it is an effective way of reducing weight, but it has a catch: very unpleasant side effects, anything from a pounding headache to profound weakness, constipation, nausea and vomiting (at least you're not hungry) to flu-like symptoms. The aim of the ketogenic diet is to starve your body of carbohydrates so that it has to use up its own stores. By not taking in any food that can be broken down into glucose, you are forcing your body to start using the reserve that is stored in the liver.

So far, we have concentrated on fat as the main store of energy in the body. In chapter 1 we mentioned that the liver also acts as an energy store. In times of starvation (or dieting), or just at times when we are using lots of energy and not taking in much (such as running marathons), the first place the body goes for easy and quick energy is the liver. We have two to three days' store of energy there. As we mentioned previously, the glucose that is stored in the liver needs to be surrounded by water and therefore this is quite a heavy source of energy compared to fat. The initial weight loss of any calorie-restriction diet is from these stores of glucose in the liver. As this energy is used up, so the water in the liver required to hold it in place is lost. The result? Lots of initial weight loss – but none of it is fat.

When trying to understand these two sources of our energy (fat and liver), imagine the human body as like a hybrid car. The hybrid car runs most of the time on its battery (in the human body it's the liver), but when the charge is running low it will switch to using the petrol tank for energy. In a similar way, in our day-to-day tasks we rely on our liver to provide our main source of energy. When the liver is running low, we are forced to switch to our other source of energy, our fat (the fuel tank). Advocates of the ketogenic diet run their hybrid energy bodies on an empty liver (a flat battery). When the liver is 'empty', the whole engine (human body) runs less efficiently and therefore it's easier to run down the tank (and lose weight).

There is another major drawback with the ketogenic diet, though – you have to remain on it for life if you are going to maintain the weight loss. This can be difficult in view of the extreme side effects and also because of the real difficulty of finding nutritious foods on the high street that have almost no carbs. When the ketogenic dieter's battery (liver glucose) is running on empty and they are 'burning' fat, they can sense this because their brain is forced to use a type of fuel called *ketone bodies* for its energy. Many followers of this diet say that this alternative brain fuel makes them feel more alert, and helps them think more clearly. Thinking fast and clearly would presumably have been an evolutionary survival advantage for our ancestors when they found themselves in an environment where food was scarce.

The ketogenic diet is so extreme in terms of its side effects, and the difficulty of finding the correct foods, that advocates have to be very determined. Presumably they get used to running on empty and love the buzz that it gives them. I personally would not advise anyone to try it. As with the majority of diets we've looked at in this book, if you lost a lot of weight dieting this way, and then reverted to eating more normally again, you would regain all of that lost weight . . . and more.

Intermittent Fasting

Popular examples of intermittent fasting are the 5:2 diet and the 16/8 diet. The 5:2 diet involves eating normally for five days of the week, and for two non-consecutive days limiting calorie intake to 500 or 600kcal. The 16/8 diet advocates eating only during an eight-hour window in the day and then consuming only tea, coffee and water for the remaining sixteen hours. This can easily be done by skipping either breakfast or a late evening meal. Both of the diets advise the need to consume healthy foods and avoid processed or fast foods.

Unlike many other types of diets, intermittent fasting, just like low-carb dieting, remains popular – which means that it probably does work for some people. Old-fashioned, conventional thinking would suggest that by fasting for long periods of time, or by cutting out meals, the number of calories consumed is decreased and therefore by using the energy in/energy out equation weight is lost. But we now know that sustained weight loss does not occur by simple calorie restriction. So how does this diet work? By decreasing the opportunities for eating, and at the same time by asking followers of the diet to avoid processed and junk foods, both the insulin profile and the omega ratio of the dieter will be improved and therefore the weight set-point will be reduced.

Vegetarian and Vegan Diets

Many advocates of vegetarian and vegan diets do so out of concern for the environment and animal welfare, but are they a reliable way of losing weight? Two of the main causes of an elevated set-point (and therefore weight gain) discussed in this book are an unnaturally high insulin profile and a relative deficiency of omega-3 compared to omega-6 essential fatty acids in our food. We know that some nomadic tribes, such as the Maasai, consume meat, blood and milk – a carnivorous diet. These tribes avoid sugar, carbohydrates and artificial oils in their diet and therefore avoid weight gain and obesity. What about the opposite to the carnivorous diet? How does avoiding animal products affect these risks?

Most vegetarians and vegans will eschew many types of processed foods because they contain animal products. This will have a positive effect on both the amount of sugar and the amount of omega-6 oils that they take in. However, most will be unaware that frying foods in vegetable oils, and consuming nuts and seeds (all very high in omega-6), will have a detrimental effect on their omega profile, particularly as a valuable source of omega-3 in fish is excluded from their diet. A pitfall, particularly for vegans (who will not consume any calories from dairy products), is to increase their consumption of bread, pasta and rice to make up their daily energy requirements. This will have detrimental effects on their insulin profiles and ultimately their weight.

In my experience most vegan and vegetarian eaters are much more aware of the quality of the food they are eating, and may be more inclined to prepare their own food. In addition, they may be more likely to avoid processed foods and fast foods. If they can avoid taking in too much sugar or refined carbohydrate (wheat), their set-point and therefore their weight will fall.

THIRTEEN

The Fat of the Land

Life Events, Hormones, Geography and Your Weight

Many of my patients (the ones who have not been obese from child-hood) will describe a particular time in their lives when they started to gain weight. Prior to this, they had not even had to think about weight regulation – it just came naturally. Then something happened, some change in their lives that resulted in a period of runaway weight gain. The life events leading to weight gain commonly include:

1. Leaving home
2. Going to college
3. Getting married
4. Doing night shifts
5. Starting a new job
6. Moving to a different country.

Once they notice the extra weight they have put on, they try and do something about it. They take the advice of their doctor or dieti-cian and go on a low-calorie diet (they might even buy the current 'miracle diet' book). Eventually (after transient weight loss) their set-point rises even further. This is when their troubles really start. Maybe after another ten or twenty years of repeated dieting and hik-ing their set-point further and further upwards they will develop leptin resistance (as described in chapter 5), leading to full-blown, uncontrollable obesity. This is the well-trodden path to the bariatric clinic that many of my patients have taken.

We know from chapters 3 and 12 why diets lead to a higher weight set-point: we are telling our bodies to prepare for future famines. But why do the other life events that my patients describe also lead to a rise in their set-points?

Let's look at these common life events and see what has changed in the environment to cause a sudden shift in the weight set-point and weight gain. Once we understand what triggers those set-point rises, we will be able to take a step further in our quest to control it.

New Horizons

First we'll look at leaving home and going to university. Why would a bigger 'fuel tank' be needed in these circumstances? Imagine young adults throughout history leaving the family tribe/home to venture off into the unknown – into the wilderness, where obtaining food might be less certain. It would be perfectly natural for the body to want a bigger fuel tank in this time of uncertainty. The elevation in the weight set-point in this case is probably brought about by the stress hormone cortisol. The same stress hormones are produced today by adolescents when they leave the family home to venture into the unfamiliar world of university.[1] Cortisol, when given as a drug to treat inflammatory conditions, has the side effect of producing a voracious appetite and food-seeking behaviour (it can also drive you slightly manic as well – which would explain the behaviour of many fresher students). And an increased appetite leads to weight gain. Higher cortisol levels therefore lead to an elevated weight set-point, which explains the 12lb (5.5kg) weight gain often seen in first-year students.[2]

Stress → higher cortisol → HIGHER WEIGHT SET-POINT

I Pronounce You . . . Hungry

What about marriage? Why should this cause significant weight gain in both women and men during the first two years, in comparison to those who don't marry?[3] Is this due to stress as well?

Although relationships and marriage can bring their fair share of stress, the significant weight gain after marriage is not down to this. In fact, studies have shown that happily married couples have significantly less of the stress hormone cortisol than single people.[4]

However, marriage is often a precursor to starting a family, and newly married couples are more likely to have a baby in the first few years after their marriage than couples who are cohabiting. So for many couples, whether they know it consciously, or sense it subconsciously, marriage is the signal to start preparing the family nest. In this day and age a couple will make practical and financial plans. They may choose to live in a home that has spare capacity in case a baby arrives, and they will have financial reserves for this possibility – many couples put off their wedding until these practicalities are in place. They will also have sorted out home and life insurance to protect them against untoward events.

However, we have not always had the luxury of guaranteed shelter and food supplies – these are modern luxuries, and changes that our genes don't understand. Throughout most of human history our nomadic ancestors would have had only one insurance policy available to them – their body's 'insurance' against famine. From an evolutionary perspective, it would make sense if the weight set-point of both the man and the woman were raised when a new baby was imminent. Both parents might need more energy reserves in the event of a pregnancy. In biological terms, this is so that the woman can carry the pregnancy safely through, even in times of food shortage, and the man has the ability to protect and feed his new family.

For women, the extra energy reserves – in the form of fat – would insure a future pregnancy against food shortage. The extra weight would also improve her fertility. From a male perspective, having a female partner carrying your baby can also pose some new challenges, ones that are common to most mammals, particularly chimpanzees and other primates like us. Having to spend time securing your territory, and possibly deterring other males from your mate, could potentially mean periods without food while your guard duties are carried out. The only sensible insurance against this would be a pre-emptive weight gain – a larger 'fuel tank' might be needed to protect and provide for your new family. The mechanisms that cause the weight set-point to increase are unclear; however, we do know that after marriage men's testosterone levels decrease and women's oestrogen levels increase.[5] A man's lower testosterone helps family stability, but also leads to weight gain; and a woman's higher oestrogen increases fertility, but also signals an increase in fat storage, particularly in the hips and breasts.[6]

After marriage, as the respective weight set-points of the couple increase (in preparation for a family), this drives their behaviour and prompts increased appetite, food-seeking and hoarding as well as producing a slower metabolism to assist the weight gain. The explanation that weight gain after marriage is because couples are 'happy' and have 'let go' is probably not true: they are biologically programmed to gain weight.*

Marriage → lower testosterone (males) → HIGHER
WEIGHT SET-POINT
Marriage → higher oestrogen (females) → HIGHER
WEIGHT SET-POINT

Night Nurses

When I was a junior doctor I spent quite a lot of time on the ward at night. I have many fond memories of the quietness of the hospital at 3 a.m. – chatting and laughing with the friendly night nurses. One of the things I remember about the nurses that chose to do night work was their size. On average, they seemed to be about 2 stone (14kg) heavier than the nurses that worked the day shift. As a junior I assumed that maybe they had made the decision for a quieter life as a night nurse *because* they were overweight or obese. It was only when I started to speak to obese patients in my clinics that I realized the night-shift work was actually *causing* the weight gain.

It is now well established that people who work on night shifts have an increased risk of heart disease, diabetes and yes . . . obesity. A recent study analysed the biological changes that occurred in volunteers who were subjected to the same sleep disruptions that night-shift workers experienced. It showed that they had decreased levels of leptin, the hormone that is the master controller of your weight.[7] Normally, leptin levels fall when we are dieting; this then causes the increased appetite and decreased metabolism required to protect the body against further weight loss. However, in the subjects who were exposed to night-shift changes in their sleeping patterns, there was no preceding calorie restriction or weight loss triggering the lower leptin levels. The cause of the leptin drop was purely a disruption in their sleeping pattern. The lower leptin levels therefore worked not to regain any weight that had been lost – but to stimulate weight gain by elevating the set-point. In addition, the scientists found that the night-shift experiment increased levels of both

---

* Some people describe putting on weight when they meet and move in with their true love. This life-change triggers the same biological nesting response as marriage – your set-point doesn't need a marriage certificate!

insulin *and* cortisol. The same stress hormone that can lead to a new student's weight gain was sky-high in night-shift workers.

Night-shift work → lower leptin → higher insulin (leptin resistance) + higher cortisol → HIGHER WEIGHT SET-POINT

I have several patients who had taken on high-flying jobs in multi-national companies. These jobs ratcheted up the amount of stress in their lives, but equally importantly involved regular long-haul flights and therefore repeated disruption of their sleeping patterns. Just like the night nurses, their weight set-point elevation corresponded to the time that they started their new high-powered job.

The Third Eye

What causes the profound metabolic changes associated with sleep disruption? Why does the weight set-point become elevated in night-workers or jet-setting business executives? It has long been known that the hormone melatonin, secreted in response to fading light and to darkness, is responsible for our day–night cycle and is the reason animals get sleepy at night and wake up at sunrise.* Recently it has become clearer that melatonin is involved not only in our sleepiness or wakefulness but also in our metabolism.

Melatonin is produced by the pineal gland (named because it resembles a miniature, 5mm pine cone) which sits just behind our eyes. The pineal gland is basically a light-sensing organ, linked by nerve impulses to the eyes. When a lack of ambient light is sensed by the gland, it will release melatonin. This is our sleepiness hormone, our 'third eye', which senses when the light is fading and readies our body for sleep. Our pineal gland therefore makes light (or a lack of it) an important neurobiological agent.

There is growing evidence that melatonin not only acts to promote sleepiness, but also has important metabolic effects, including an increased sensitivity to leptin and a decrease in cortisol.[8] If we are sensitive to leptin, our weight set-point should stabilize. However, in circumstances where there is a decrease in pineal stimulation, caused by working through the night and trying to sleep during the day (a deficiency of darkness), melatonin levels will be reduced,

---

* This applies to diurnal animals – those that are awake and active in the daytime.

leading to reduced leptin sensitivity and – you guessed it – a higher weight set-point.

**Decreased melatonin → leptin resistance +
higher cortisol → HIGHER WEIGHT SET-POINT**

The actions of the pineal gland, and melatonin's effect on leptin, cortisol and metabolism, are still being researched. Some scientists have speculated that the *lack* of darkness in our neon-lit cities could be having a profound effect on the metabolism of the citizens living in them – with melatonin deficiency contributing to diabetes and obesity.

Passport to a New Weight

The final, but increasingly common, life event that can precipitate sudden weight gain (which I see in patients who have never had problems with their weight before) is migration. Now that we have the mechanical means to 'fly', humans as a species commonly migrate to live in places far away from home. However, unlike birds, whose decision to migrate is dictated by the seasons, the human decision to migrate tends to be based on economic or family circumstances. A bird will fly towards an environment that will suit its future health – they are always flying towards summer.

Moving to a different country clearly has an effect on the environmental clues that your brain is searching for in order to calculate the safest quantity of energy reserves for the future. Are you moving into an environment where a famine or a food shortage is going to be more common? Are there signals that a long winter is approaching?

I have seen many patients who migrated from Asia or Africa to Britain and started to gain weight. Some of them gained weight immediately, others after a few months or years. The weight change usually started when they switched from their traditional foods to a more Western diet. Similarly, many patients who travelled from the UK to the USA put weight on: typically their weight increased and then settled at their new 'American weight'. Interestingly, in the patients that made the move back from the US to the UK, their weight settled back down to their 'UK weight'. Americans who migrate to Dubai lose weight, but the British who travel to work in Dubai gain weight. And even *within* countries weight change can suddenly occur: Indians who had previously lived in a rural environment and then migrated to work in the cities complain of weight gain.

In most cases, people who move to an environment where there is a Western diet will be at risk of an increase in their weight set-point. What I find interesting is that there seems to be a hierarchy in the areas that have Western food available (probably determined by the omega profile of a country's food). America is at the top: people who migrate there will increase their weight and people who leave will decrease their weight set-points. Next is the UAE, followed by northern European countries, then southern Europe, then big cities in the developing world, like Mumbai or Delhi.

We discovered in chapter 9 that most populations which are exposed to a Western diet will have a significant deficiency of the essential fatty acid omega-3 and a massive excess of omega-6. This is largely because fast foods and all processed foods have low omega-3 and high omega-6. A population that is exposed to these types of food will develop cell membrane changes that mirror the ratios of these fatty acids in the food supply. These changes can raise the weight set-point in those people who are genetically susceptible (the Labradors, not the greyhounds*), leading to weight gain.[9] In others, the cell membrane changes that the Western diet causes may not lead to weight gain but may trigger other modern diseases such as arthritis and heart disease.

The concept that the omega-3 to omega-6 ratio of a country's food is mirrored by the ratio in the cell membranes of their population fits in with patients' observations. Patients suggest that when they move to a new country, their set-points alter, according to their exposure to the Western diet. Although people in Europe, the USA and the UAE all consume a 'Western diet', the constituents of the diet will differ: for instance, in the USA 70 per cent of the calories consumed are from processed foods.[10] Anyone who has travelled to the US will know how difficult it is to eat healthily. In the UK, 50 per cent of the food consumed is processed. If you travel to other parts of Europe the percentage is lower: 46 per cent in Germany, 35 per cent in Austria, 20 per cent in Slovakia and 13 per cent in both Greece and Italy.[11]

---

* Imagine the difference in obesity genes between a Labrador, which has a common genetic mutation that causes a susceptibility to obesity, and a greyhound. Give them both processed dog food (with the omega-3 taken out) and the Labrador will invariably become overweight, but the greyhound can eat as much as it likes and it will not put on excessive weight. These differences in genetic sensitivities to obesity exist just as dramatically in humans.

As the immigrant assimilates the omega fatty acids of the new country's food into their body, the omega-3 to omega-6 ratio in the food and the omega ratio of their cell membranes become identical. If the new country's food has a higher omega-6 to omega-3 ratio (i.e. more processed and fast food is available than in the country of origin), then the immigrant's weight set-point will rise. If the new country's omega-6 to omega-3 ratio is lower than their country of origin (less processed food), then that person's weight set-point will decrease. Our patients describe their weight settling either upwards or downwards depending on the quality of the Western food in their new environment. This usually takes several months as the dietary changes become embedded into cellular metabolism.

High processed-food intake = high omega-6 to omega-3 ratio in food → high omega-6 to omega-3 ratio in cell membrane → higher insulin resistance + more leptin resistance → HIGHER WEIGHT SET-POINT

The Microbiome and Weight Loss

If our external environment has such a profound effect on our weight set-point, what about our internal environment? There has recently been a lot of interest in the microbiome, those billions of bacteria (as well as fungi and viruses) living within our guts. Could these affect whether we gain or lose weight easily?

Since 2014 scores of articles have been published in scientific journals linking changes in the make-up of our gut bacteria to obesity. However, the study of the microbiome is very new and we need to approach this new science cautiously and with an open mind. We certainly know that our gut bacteria can have a profound effect on our health, as we know only too well if we suffer a bout of gastroenteritis. The association between a single bacterium (E. coli, for example) and the symptoms of gastroenteritis is obvious. But what if we decided to look at the relationship between the relative populations of 1,000 different bacterial species in our guts and link it to obesity? We suddenly have a large tangle of bacterial data and an outcome measure (our weight) which is not binary (there is no definitive answer as there is with the symptoms of gastroenteritis). Mix into the pot the fact that many studies have been discredited because of contamination and things start to get confusing.

In addition, gut bacteria are affected by the types of diet that we eat. If we eat a typical Western diet that is low in fibre, then the microbiome in the gut will have less diversity. Therefore, to conclude that changes in gut bacteria cause obesity you need to somehow separate the two facts that a Western diet causes obesity, and also that a Western diet causes changes in the gut bacteria. An impossible task!

So why has so much scientific and media interest been focused on a possible link between the microbiome and obesity? A trip to any health-food shop may clarify things. Large sections of these stores are now dedicated to the sale of probiotics – capsules containing bacteria that scientists tell us are *good bacteria*. In 2016 the probiotics industry was valued at $4 billion. By 2022 the microbiome market, encompassing research and development, plus treatments in the form of probiotics, prebiotics and medical foods, is projected to be valued at $6.9 billion. This new industry, fuelled by self-funded research and the interest of our inquisitive media, is growing at over 9 per cent per year.

I am sure the microbiome will play an important role in our understanding of many diseases in the future, but at present there is no reliable evidence that it has a direct causative effect on our weight set-point.

The Only Way to Lose Weight: Lower Your Set-Point

I hope that it is clear now that sustained weight loss can *only* be achieved by lifelong eating and lifestyle adjustments – changes that will lower your weight set-point and give you a better quality of life.

Now that you understand the basic principles of the weight set-point (how it goes up or down in response to signals from your diet and environment), we can start to plan the changes that will suit you as a person.

In the final part of this book we will examine in more detail *how* you can lower your weight set-point by altering:

1. Your environment and psychological health
2. Your food and eating habits
3. Your activity and lifestyle.

Throughout this book we have discussed how sugar and highly refined carbohydrates (such as bread and pasta) give us a great feeling. We are hard-wired for sugar to give us a similar euphoric feeling to that of a drug addict after a hit. It's not our fault, that's just how we evolved. It is for this reason that many people suffer with a kind of addiction to the hedonic, pleasurable feeling that foods containing sugar and wheat give us. These food addictions can be difficult to conquer and it is for this reason that I have included a section on psychological techniques that can help you, contributed by a bariatric psychologist.

PART THREE

Blueprint for a Healthier Weight

The Secret to Lasting Weight Loss

FOURTEEN

Prepare to Do It Yourself

Preparing Your Home and Mind

One of my biggest nightmares, one of the most frustrating parts of my life, is DIY. Despite being a surgeon, and therefore being good at using my hands, I seem to have great difficulty with even the simplest of DIY jobs. A weekend trip to IKEA sends me into a cold sweat. In the shop the furniture looks Scandinavian, stylish and simple, but once I make the mistake of buying it – and empty the contents of the box (hundreds of different nails, screws and washers; scores of different shapes and sizes of wood) – I am reminded why I don't visit regularly (despite the tasty canteen meatballs). After several hours on my knees, or adopting various different Twister-type positions to fix screws into annoyingly difficult places, I realize that I have not read the instructions properly – and have to start again . . . When I finally finish constructing the new cabinet I find that, suspiciously, there are still some screws left over that should have been used. The function of these screws becomes apparent a couple of years down the line when the cabinet door dislodges . . . and just as I try to fix it back on . . . a leg falls off.

But when I think back to my DIY disasters I realize the simple mistakes I keep on making: not reading the instructions properly, not taking the time to prepare, being rushed and, most importantly, not having realistic expectations (i.e. thinking a complex cabinet can be knocked up quickly).

Prepare to Reset

In this chapter, we are going to look at the necessary preparations to change your weight set-point. As with a complex DIY task, unless you prepare the groundwork for the job ahead, it is likely to end in failure or disappointment. The aim is to alter small parts of your environment and also the way you live, your habits. These changes will be sensed by your weight-control centre – and your weight set-point and then your weight will drop.

What you need:

1. Realistic expectations
2. An understanding of how to fix the problem
3. A prepared home environment
4. Time.

Realistic Expectations

A crucial part of the preparation for losing weight, and keeping weight off in the long term, is to have realistic expectations. The success of your efforts to lose weight will be determined by how honest you are with yourself as to what can sensibly be achieved.

If you have gained a lot of weight during your adult life, then it is unlikely that, even with optimal lifestyle changes, you will be able to reach the weight you were at the age of eighteen as your body will be biologically different. If you have very strong obesity genes, if you are from a 'big' family, then again it is unlikely that you will become slim. An expectation to lose some weight and be healthier and happier is much more achievable.

As we have learned, everyone is different when it comes to their propensity to gain weight in a particular environment. And the same can be said when we put in place habits and changes to our environment that will lower our weight set-point. The main expectations that you should have when you make these changes are that, yes, you will lose weight, but, just as importantly, you will be much healthier and will live a longer and happier life as a result. Try and focus on your improved health and happiness just as much as your improved waistline when you are visualizing your life in the future. Don't get frustrated if you don't reach your weight-loss target. It may be that it was unrealistic. Or it may be that you have not allowed enough time for the changes to work through your body.

Unlike a short-term dietary fix, this programme will build momentum and become more and more effective as time goes by, as you embrace a lifestyle that suits you and insulate your body from the dangers out there in the environment.

Some patients do come to me for bariatric surgery with unrealistic expectations, however. Maybe they weigh 120kg (19 stone), but they say they will only be happy if they can reduce their weight to 60kg (9 stone 6lb) – they want to be skinny, even though they are middle-aged and have never been remotely slim in their lives. If you have the surgery, I tell them, your weight will reset to 80–85kg (about 13 stone) – not 60kg. If they go ahead with unrealistic expectations, then even if the surgery is successful, even if they reach the anticipated 80kg, they will be disappointed and may see the procedure as a failure. So realistic and open expectations are the key to success.

Understand How to Fix the Problem

If you are serious about weight loss you should not skip immediately to this part of the book and start the programme. First, you need to understand how your weight is controlled and how your body interacts with your environment for it to work. *It is crucial to understand the concept of the weight set-point – this will be the key to your success.* In the same way that you need to study and understand the instruction manual before you begin to construct a complicated piece of furniture – otherwise it might end in frustration – so you need to understand this new concept of weight regulation before starting the programme. If you feel it is not clear to you yet, go back to Parts One and Two to refresh your understanding of the problem and the solution.

Prepare Your Home Environment

We know that when people move from one country to another they take on the imprint of that country's food within their bodies. Even if we try to eat in a relatively healthy way, the country's food and food culture will still prevail. The main purpose of this book is to enable you to sustain weight loss by decreasing your set-point. This can only be done by changing the environmental signals that you are receiving. None of us really want our bodies to experience the full Western food (and stress) experience as this is what has poisoned it and led to the metabolic chaos, resulting in excess weight and obesity.

If you want sustained weight loss, you cannot rely on food products that are 'industrialized' or processed – they contain too much sugar, wheat and vegetable oil. The best way to be sure that you are eating food that is good for you is to buy fresh ingredients and *cook them*. This is a very important point. Success will only come when you understand which foods to eat. And it will only be sustained if that food tastes great and if you look forward to eating it – and cooking is the best way to achieve this.

Remember: The Master Chef is You

Even if you have never tried cooking before, it is not too late to learn. It will enrich your life – weight loss with no hardship, just great food. If you can't cook well you may want to consider taking lessons or learning from a friend or relative. Another possibility is to get free cooking lessons online – Jamie Oliver is always there with a load of fresh wholesome ingredients to teach you. Another option to improve your cooking skills is to try a food delivery company such as Gousto or HelloFresh. They will deliver a box of fresh ingredients right to your doorstep. The box also has simple instructions on how to cook the food – and you end up with a filling and nutritious restaurant-quality meal. If you have a family this can also be a social activity: you can take it in turns to choose your meal and cook it.

Cooking should be a pleasurable experience – remember, this is what made us human in the first place. Once you learn a variety of ways to cook, you may find that it becomes one of the highlights of your day – the actions and concentration that cooking requires will naturally relax you. After a while, cooking will have become a valuable part of your life, and maybe even something to pass on to the next generation.

To make your cooking experience great, in addition to having a standard set of pots, pan and dishes, I would advise investing in a new set of knives, an old-fashioned knife sharpener, a hefty chopping board and a food blender. You may need to clear out your fridge and start again to fill it (and the freezer) up with fresh food. Set up your radio or Wi-Fi speaker and enjoy the whole relaxing experience – your mind will be focused on your food (and the music) and all troubles and worries will fade away; you'll also be able to show off your new-found skills to family and friends.

Away from Your Frying Pan

The other preparations to make before getting started are to your home environment – especially in the lounge and bedroom. Part of our plan will encourage you to sleep more, and in order to do this you need low lighting in the house for the hour or so before bed. See if you can change your bulbs to low wattage, invest in low table lamps or get dimmer switches. You may even want to choose a book to help with that relaxing time in bed.

Oh, and a final change in your home environment – please throw away your scales! The weight loss and health will come; don't become stressed and obsessed by it and don't be tempted to force it.

Reduce Family or Work Stress

We have learned that our environment can affect our weight set-point. External stresses and anxieties in our lives can also influence our cortisol levels. This will have an impact on where our bodies want our weight to be – how big a 'fuel tank' might be needed. If you are unduly stressed, then your metabolism acts like that of an injured animal – the cortisol message means you're not going to let go of energy stores easily.

There may be factors in your life that are difficult to control, but that are causing you an undue amount of stress and therefore affecting your cortisol levels and your weight (cortisol → raised set-point). Before you start the programme, you will need to take stock and consider what these factors could be. Are you in a job that is particularly stressful? Are there family, or relationship, issues that are causing stress at home? Is the commute to work too much? Are you incessantly worried about money? These factors are just as important as your diet. Unless you address these issues you may find your set-point more difficult to shift downwards. Remember, if all else fails, sleep, exercise, music, massage, dancing and laughter all help to lower cortisol levels.

Time to Start

You are nearly ready. You have realistic expectations that you are going to become much healthier and slimmer; you have read this book and understand how to regulate your weight and you have prepared your kitchen and home environment. The final, and perhaps the most important, ingredient is *your time*. To make lifestyle changes you will need to find the time to shop for good fresh foods and then the time to cook them, as well as finding time to be active and extra time for rest.

If you can make time to look after yourself, then everything will be easier. Remember that you are building a new body and a new life; your time investment is needed to do the job properly. If you are in a busy job or you have lots of family or other commitments, then you need to stand back and take stock of your life. Look to see how you can find this precious time. It may be that your current down-time is used mindlessly watching Netflix or scrolling through social media feeds. Think seriously about your lifestyle and your daily routine to find this time – only you can do it. I would estimate that you need to find one to two hours extra per day by dropping or curtailing current unproductive activities.

It Will Only Work If You Enjoy It

The crucial part of lowering your weight set-point is in the food and lifestyle changes that will be outlined in the next two chapters. If these changes improve your life, if they make you happier, then it is likely that you will continue with them – with the bonus that your weight will have reset, permanently. This is the opposite approach to the short-term diets in which the changes necessary for weight loss make you feel unfulfilled, unhappy and hungry – and are therefore not sustainable.

Prepare Your Mind

One of the essential changes that you must make to reduce your weight set-point is to normalize your insulin profile: this means no more sugar surges. We will discuss this in more detail in the next chapter. But you need to be mentally prepared for this change in your eating habits. As we have learned, sugar and highly refined carbohydrates (like flour) directly stimulate the reward pathways in your brain. This results in a surge of the brain hormone dopamine; it makes us feel good. It is naturally released after we consume foods containing sugar, and after sex. However, it uses the same pathway as drugs such as alcohol, nicotine, cocaine or even heroin. Because the reward is so powerful, we can get hooked on it and get addicted to the substance that triggers those feelings – whether it be drugs, sex . . . or sugar. You may have a sugar addiction, partial or full-blown. In either case, when you limit your exposure to sugar, you might find yourself going 'cold turkey'. You may experience headaches, muscle aches, fatigue and poor sleep – until you can satisfy your extreme craving for sugar to stop the withdrawal symptoms and get another shot of dopamine.

You need to be mentally prepared for these short-term side effects of your change of diet. Just like the preparations you make in your home environment, you need to prepare your mind for the changes to come. You need to adapt to a healthier way of eating – but because of the dopamine surges that sugar gives us, and the addictive nature of these rewarding feelings, you may need extra help.

It is essential to have a clinical psychologist as part of any bariatric surgery team. Most patients with severe obesity will have developed an addiction to sugar and sugary foods, and as a result will also be suffering from a leptin deficiency, so that their brains will be receiving hunger signals constantly (as we covered in chapter 5). This is the reason most of these patients are compelled (by their hormones) to binge-eat, to eat to excess, usually in secret, while simultaneously experiencing a lack of control and feelings of guilt afterwards. Their problems are exacerbated because, by the very nature of binge-eating and the compulsion to consume as much energy-giving food as possible, it is highly likely that they will also become hooked on sugar. The psychologist plays an integral part in helping obese patients through the process of mentally readapting after bariatric surgery, when sugar is suddenly off the menu.

A Guide to Mindfulness and Mindful Eating

My friend and colleague Jackie Doyle, Lead Clinical Psychologist in my department at University College London Hospital (UCLH), has offered some useful advice on how to be mentally prepared for the changes in diet, and how these changes might affect you – as well as suggesting some practical coping strategies.

SECTION 1: MINDFUL MEDITATION

The best way I know to help manage stress is through mindfulness meditation. Numerous studies have shown that practising mindfulness improves emotional wellbeing and quality of life. There is also growing evidence that it can have a direct biological effect on stress hormones such as cortisol.

Mindfulness allows us to be fully present, aware of where we are, what we are doing and how we are feeling and thinking. It is in contrast to the state of being on autopilot, which will be familiar to all of us. This morning I was sitting at my desk and I realized that the drilling that was going on in the road outside had stopped. The sudden quiet was a real relief to me, which was strange because I hadn't really been aware of the effects of this noise on my concentration levels. For all of us, sensations, emotions and thought processes can be occurring in the background without us really noticing, as we go about our daily lives. For example, have you ever had an experience where you suddenly become angry about something, but later feel you had over-reacted? Again, these 'out of the blue' reactions often occur because we have not been aware of the 'back story', a whole host of narratives, physical sensations and emotions occurring in the background of our experience. With mindfulness, we can become more aware of these background experiences and through this awareness learn to respond to difficulties in a skilful way, rather than reacting on autopilot.

Please don't be alarmed: learning to meditate does not mean heading off into the hills of Tibet, and indeed you don't even need to channel your inner Buddhist (unless of course you want to). There are many ways that people can learn mindfulness, through courses in person or online and through books or computer apps. Professor Mark Williams, recently retired Professor of Clinical Psychology at the University of Oxford, has developed a series of resources under the blanket title of 'Finding Peace in a Frantic World'. More information on this can be found at https://franticworld.com.

SECTION 2: MINDFUL EATING

There are so many myths and misunderstandings about food and weight regulation. Often the people I meet have lost confidence in knowing how to eat, and what to eat. Research has shown that if babies are given access to a wide range of unprocessed food, they will eat a balanced diet over a period of about a week. However, for a variety of reasons we seem to have lost this natural intuition about eating.

Eating has become one of the things that we do very mindlessly. It is now very much a feature of Western culture to eat on the run, eat at our desks or in front of the TV. It has come to the point where many people say that they feel uncomfortable eating on their own if they are not also doing something else, e.g. checking their phones or reading. One of the downsides of this is that we miss valuable information about our eating habits.

In my clinical work, I run mindful-eating groups or seminars. At a certain point in proceedings, I hand out a small piece of chocolate for people to try. This is often greeted with laughter, surprise and fear.

People are often shocked that I would encourage chocolate-eating, but go on to enjoy the 'naughty' element of this exercise. Some people, however, refuse to take the chocolate because they fear that it will open the floodgates to a full-on binge.

Below you'll find a short exercise that you can try for yourselves. When trying this at home, I encourage people to choose a chocolate that they like, have eaten before and would often choose to eat. However, if it feels safer to choose a different food, this is also fine. The main thing is to try the experiment rather than just read about it.

MINDFUL EATING EXERCISE

Begin sitting down, ideally in a place where you are unlikely to be disturbed for five minutes. Have the chocolate unwrapped in front of you. Start off by really looking at the chocolate as if you have never seen it before. What do you notice; is it as you expected it to be? Now note what is going on in your body. Perhaps you are aware that your heart is starting to quicken or your mouth has begun to water?

Take the chocolate and smell it. What do you notice now? Does it smell as you expected it to? Does it smell different in one nostril compared to the other? Again, note what is going on in your body and perhaps any thoughts that you are having. Ask yourself, how much do I want this chocolate on a scale of 1–10 (10 = I very, very, very much want the chocolate!). Now take a small bite of the chocolate and rest it on your tongue, without chewing, and put the rest of the chocolate down for a moment. Slowly begin to chew, noticing all the sensations in your mouth, stomach and the rest of the body before you swallow. As you swallow, see if you can pay attention to the sensations as the chocolate goes down your throat into your stomach. Once the chocolate has gone, take note of what is happening in the rest of your body and any thoughts you are having. How satisfying was the chocolate on a scale of 1–10 (10 = extreme satisfaction)? How much do you want more of the chocolate on a scale of 1–10?

Now repeat this sequence. Take a small bite, rest it on your tongue, without chewing, and put the rest of the chocolate down for a moment. Slowly begin to chew, noticing all of the sensations in your mouth before you swallow. As you swallow, see if you can pay attention to the sensations as the chocolate goes down the throat into the stomach. Once the chocolate has gone, take note of what is happening in the rest of the body and any thoughts you are having. How satisfying was the chocolate this time on a scale of 1 to 10 (10 = extreme satisfaction). How much do you want more of the chocolate on a scale of 1–10?

Finally, repeat the sequence one last time. Once again take a small bite and rest it on your tongue without chewing, placing the rest of the chocolate down for a moment. Slowly begin to chew, noticing all of the sensations in your mouth before you swallow. As you swallow, see if you can pay attention to the sensations as the chocolate goes down the throat into the stomach. Once the chocolate has gone, take note of what is happening in the rest of the body and any thoughts you are having. How satisfying was the chocolate this time on a scale of 1 to 10 (10 = extreme satisfaction). How much do you want more of the chocolate on a scale of 1–10?

After you have completed this exercise, reflect on what you noticed. How might these discoveries help you take care of yourself?

This exercise is often a real eye-opener. Some people discover that they don't actually enjoy the chocolate as much as they thought they did. It smelled funny or it tasted too sweet or they didn't like the sensation at the back of the throat. Others say that the experience of eating slowly gave them as much satisfaction from three small bites as they would get from a whole bar of chocolate. This may sound crazy, but it is really worth testing it out. Others say that the exercise piques their desire to eat more chocolate, in which case I encourage them to do this, but again in this slow mindful manner, and see what they discover. Most people find that they are satisfied by much less than a typical portion size.

It would be a shame if readers were to embark upon this new adventure and miss the pleasures that food can bring or indeed miss the fact that food they once thought of as enjoyable actually isn't. In the words of Jan Chozen Bays, American paediatrician and mindfulness expert, 'Mindfulness is the best seasoning.' For more on mindful eating, see the website (mindful eating, conscious living, https://me-cl.com).

SECTION 3: MANAGING CRAVINGS

Many people report that when they start to eat well, they experience fewer cravings. However, eating is not just in response to physical hunger or a result of hormonal changes in the body. There are times when certain foods seem to be calling us! The biscuits bought for the children or grandchildren appear to shout 'come and get me' from the cupboard. The chocolate in the fridge transmits thoughts such as 'one won't hurt', 'you deserve a treat, you have had a hard day.'

It is important not to underestimate the intensity of cravings. I often ask people to describe how their bodies feel when they are in the grip of an intense craving. They say that they feel edgy, agitated and restless; some say it is like being itchy all over. More importantly, in this state people say that it is very difficult to reason with themselves, to talk back to the chocolate or biscuits that are calling them. This is in fact an example of the autonomic stress response, when the body is charged, ready to act, and the capacity to think is reduced. This is an important physiological response in times of danger, but not so useful when this system is turned on in relation to food. If this is a regular experience, it is important to come up with a plan ahead of time, before the craving strikes, so that it can be implemented when it does. There are a variety of ways that I suggest people manage cravings.

- **Get moving** – as mentioned above there is a lot of physical agitation associated with cravings. Moving can effectively 'burn up' this extra energy. Some of my clients turn the music up loud, dance, march on the spot or get involved in other intense physical activities. It is not simply about distraction; it is about directing this extra energy effectively.
- **Surfing the urge** – what goes up, must come down! People often believe that the urge or craving will just keep building and building. However, when we allow ourselves to really observe what happens when the cravings emerge, we often notice they build in intensity and then something happens and the urge disappears (until the cycle potentially repeats itself). I often get people to write down a self-coaching mantra, a statement that sounds convincing to them, something that they would like to say to themselves when the cravings bite, for example, 'Don't panic, this will go away', and practise using this each time they feel driven to eat in a way that doesn't fit with their overall goals. The important thing here is that this mantra is prepared in advance as our capacity to 'think straight' in the moment may be somewhat impaired.

- **And breathe . . . !!!** Cravings arise from a high-energy state, which we can lower by simply remembering to breathe again. I have added an exercise for a three-part breathing space below, which, if used over time, can help us respond more skilfully to all sorts of difficulties and uncomfortable experiences.

HOW CAN BREATHING MINDFULLY HELP?

The breathing space provides a way to step out of autopilot mode and reconnect with the present moment. The aim is not necessarily to relax; you may find that you are not relaxed at the beginning or at the end. However, giving yourself this space can help you decide how to respond thoughtfully to any difficulties or discomfort that might be around. The three parts are as follows:

1. Awareness
Begin by sitting comfortably, feet firmly on the floor and, if it is possible, close your eyes. Start by noticing the sensation of being held by the chair and the points of contact and pressure. Now ask yourself, 'What am I experiencing right now, which thoughts are going through my mind (pause), which emotions or feelings are around (pause) and which physical sensations are here (pause)?' You do not have to change anything, just register what you are feeling right now, even if you don't like it.

2. Gathering
Move your attention to the place where you feel the breath moving in and out of your body most vividly. Try to give your full attention to each in-breath and to each out-breath as they follow, one after the other. This focus on the breath will bring you into the present and help you tune into a state of awareness and stillness. (You may decide to count a certain number of breaths, with each in-breath and out-breath as one count.)

3. Expanding
Now move your attention away from the breath and bring into your awareness the entire body. See if it is possible to notice and hold a sense of the whole body from the soles of the feet to the crown of the head,

including a sense of the arms and hands, the trunk of the body, the head, neck and face.[1]

An audio download of the three-minute breathing space can be found at https://franticworld.com.

Of course, it takes time to create new habits. Initially, trying one of these techniques might feel like trying to write with your non-dominant hand, or ignoring rather than scratching an itch. However, over time these techniques do work and are worth trying out!

Dr Jacqueline Doyle is the Lead Clinical Psychologist, UCLH Centre for Weight Management, Metabolic and Endocrine Surgery, University College London Hospitals NHS Trust, London, and Director of Living Well Psychology Ltd.

FIFTEEN

Eat More, Rest More

Lowering Insulin and Cortisol

Eating more and moving less may sound counter-intuitive to weight loss. Surely, this will lead to gaining weight? This is what our energy in vs energy out weight-loss equation tells us.

But, as we have learned, energy in vs energy out (the amount of food that you take in and the amount of energy that you expend) is outside our conscious long-term control.

We know, from what dieters tell us, that eating less (by dieting) and moving more (by flogging yourself in the gym) do indeed lead to short-term weight loss. However, this is rapidly followed by metabolic adaptation by the body (slowing the metabolism and increasing hunger hormones), and by weight gain to an often higher weight than before the diet. Your body has been scared into a higher weight set-point.

So we are going to try something different. Instead of dieting, we are going to change the environmental signals that our body is picking up. These signals will reset your weight set-point at a lower level and then the hormonal and metabolic signals will drive your weight downwards (towards the lowered set-point). As a consequence, you will notice a natural decrease in your appetite and a natural increase in your metabolism – making you feel more energetic and alive.

In the programme that I am about to set out there is simple step-by-step guidance to move that set-point down. Each step will gradually help you lower your set-point. It will be like a cross-country run: the steps at the start are quite easy, the middle section is enjoyable, but the final steps may be more difficult. Everyone has different goals in weight regulation – some people want to lose 5kg (10lb) and others may want to lose 20kg (over 3 stone). And every person has different genes, meaning weight loss is more difficult for some than others.

If you reach your target weight before the end of the course (after only one or two steps for instance) you can stop there if you want to – you don't have to run the whole race. But this is really a lifelong plan. The steps should be integrated into your daily routine, your normal way of living. This is the only way of not only losing the weight but keeping it off for good.

STEP 1 – EAT MORE

The first step will focus on decreasing your average daily insulin levels – but at the same time you can eat really well. As we learned in chapter 10, insulin levels will drive your weight set-point up or down. We also learned that insulin levels are controlled by the types of food that we eat. The big drivers of insulin spikes in our current Western diet are sugar, wheat and corn. We will replace these with more natural foods that are less damaging to our metabolism.

In this part of our plan we will be flooding our body with lots of tasty and nutritious foods. This will not only provide great sources of extra vitamins (particularly B vitamins that optimize metabolism), but also calm down our body's cortisol levels. We are not going to shock our bodies by any type of calorie restriction.

This part of the plan is designed to help you come away from any addiction to sugary foods and at the same time to encourage you to cook good and delicious foods. We want this to become your new routine before progressing to the rest of the course.

The simple instructions of this step are:

1. Eat three meals per day
2. Have a high-fat/protein and low-carb breakfast
3. Cook/prepare your own food
4. Avoid sugar, wheat, corn, and fruit juices
5. Snack if required.

Clearing out the Pantry

The final preparation for this step of our programme is to make sure that you have cleared the food cupboards and fridge of the foods that you are going to avoid – including snacks that contain sugar – and filled it with more nutritious alternatives. There should be no bread in the house; you can probably remove the bread bin as you won't need this any more. Perhaps put a blender (for soup) in its place.

Any foods containing wheat should go – these include biscuits, cakes, crackers and many types of processed foods. Sugary snacks and confectionery should not be available. Replace the 'treat' box with a visible fresh fruit bowl. Pre-packed fruit juices and dried fruit should also be avoided as these will give you a sudden sugar bomb.

Have healthy snacks in the fridge: meat, cheese, boiled eggs, yogurt, full-fat milk. Vegan options would include hummus or salsa dips; sliced vegetables, avocados, rice cakes, baked vegetable chips, and dried coconut or even dark (non-sugar) chocolate. Fresh fruit is OK, but try to limit yourself to a maximum of two pieces of fruit per day. Have a book for recipes and cooking notes handy. Remember you are going to be cooking a wide variety of foods, so make sure you have a full range of cooking spices prominent and have herb plants growing and available on the windowsill. It's important to make these changes to your kitchen. If unhealthy 'sugar-rush' food is in the home, it's more likely that you will eat it.

If you live with children, then you should get them on board with this new food environment. You cannot force them to eat home-cooked foods outside the house, but as they see you looking happier, healthier and slimmer, they may slowly be converted.

The Full English

At this stage we want to even out the glucose fluctuations in your bloodstream – so it's important to start this process in the morning. We know that when you eat a high-carbohydrate breakfast, you will experience a mid-morning fall in your blood-sugar levels which will drive you to seek more sugar and carbs. The high-carb and low-fat breakfast sets you up for that 'sugar roller-coaster' ride, resulting in high average levels of insulin throughout the day, driving up your set-point. We want to reverse this.

In this first step of the programme I want you to avoid the type of low-fat, supposedly healthy breakfast that we have been told to eat by nutrition experts for the last thirty years. Avoid toast, cereals that contain sugar, wheat or corn, yogurt that contains lots of sugar (if it is labelled 'low fat', this is code for 'high sugar'), and avoid orange juice (or any fruit juice). A traditional full English breakfast,* or even

---

* Do not be scared of cholesterol or saturated fats – we have discussed the flawed research in this area at length in chapter 8. Natural saturated fats (not artificial polyunsaturated vegetable oils) do not make you fat and have less risk than sugar for heart disease. Replacement of these fats in our diet by oils and sugar in the 1980s

avocado, salmon and eggs, is fine if you have the time. The one proviso is *not* to have bread with your eggs, bacon, sausage and tomato. It may seem unusual to have a full English on the first morning of your new food diet. But you will see that this time will be different (you may even gain a small amount of weight before your body adapts to lower insulin levels and settles at your new lower set-point).

Other acceptable low-carb but high-fat protein breakfasts could include eggs (boiled, fried, poached or scrambled) on their own, a continental-type cold breakfast of meats, cheeses, yogurt and olives, any type of fish or full fat milk, oat porridge (with salt or a small amount of honey for flavour). It's probably best to avoid fresh fruit for breakfast due to its natural sugar content – save this for later if you require a snack. You should drink water, milk, tea or coffee (a full-fat latte is also OK).

You will notice that after this type of breakfast you are well set up for the day. You will not crave mid-morning carb snacks and you will not be ravenous at lunchtime.

Prepare for the Food Desert

Lunch choices can be difficult, particularly if you work in a town or city. Remember, despite our city shops overflowing with food, we still live in a virtual 'real food' desert. Foods that do not contain added sugar or wheat can be difficult to find; many of the food labels tell you 'low fat' (lots of sugar) or 'no added sugar' (lots of fruit sugar). If you have not cooked or prepared your own lunch, you might be tempted by these types of food, or even hedonic fast food. It is for this reason that this part of the plan tries to get you into the routine of planning your lunch and taking it with you if you are out and about.

The easiest option is to prepare your lunch the day before. It could be a home-made soup, a chickpea or pasta salad, egg-fried rice with cauliflower, mushroom, meat or fish etc. The choices are countless and just need some thought and imagination: try and get into the habit of using fresh vegetables, dairy, meat or fish to prepare a tasty and nutritious lunch. If you are a meat-eater, then regularly roasting a beef or lamb joint can provide for many lunches and snacks.

---

precipitated the current obesity epidemic. There is an additional scientific section on cholesterol in Appendix 1 of this book.

Evening Meal

You are going to spend your new-found time enjoying cooking the evening meal. You can use any ingredients so long as there is no excess of wheat, sugar or corn. This part of the programme is *not* designed to starve you of carbohydrates or send you into starvation shut-down mode – potatoes and rice are OK. If you usually had a dessert course at the end of your meal prior to this programme, why don't you consider replacing this with a starter course such as home-made soup?

If you live with your family, try and get them around the table to share the food. It's best to put all the food dishes in the middle of the table so that people can help themselves and take more if they need to. Enjoy your food, savour it. Try and make mealtimes what they should be, and what, historically, they used to be: enjoyable, sociable occasions and an integral part of your life.

STEP 2 – GET MORE SLEEP

This sounds like an easy step to take. 'No problem with this one,' I sense you saying to yourself . . . but what we want you to do is improve the *quality* of your rest, and that may not be as easy as it seems.

We all fall into mindless habits in our down-time. One of the most common is sitting in front of the TV – switching it on and switching ourselves off. I am as guilty as anyone else in this respect after a stressful, busy day. The average TV viewing time per day in the UK is reported to be three hours and twelve minutes.* Total 'audio-visual' screen time is estimated at five hours, and with the advent of on-demand channels like Netflix and YouTube the time that we spend looking at a screen in the evening may be increasing even more.

The problem with our unproductive screen time is that it can encroach on our sleep time. In addition, if we have only just switched off a bright screen when we finally do make it to bed, it can make falling asleep more difficult. As we found out in chapter 13, our 'third eye', the pineal gland, senses when light is fading and stimulates the secretion of the sleep hormone melatonin. A lack of melatonin in night-shift workers is thought to explain their weight

---

* Ofcom *Media Nations 2019* report for the UK, published 7 August 2019.

gain. It has an integral effect on high cortisol levels and leptin resistance – causing an increase in the weight set-point.

A lack of sleep has also been proved to cause increased ghrelin (the appetite hormone). Not only are you leptin-resistant and metabolically sluggish, but you also have a sharper appetite and cravings for high-energy foods. In studies comparing groups of people who slept 4½ hours or 8½ hours, those with a sleep deficiency consumed 300kcal of energy more than those who got a good night's sleep. In similar experiments a lack of sleep can cause average blood sugar levels to increase, leading to pre-diabetes.[1]

From an evolutionary perspective, sleep deprivation may have been a factor long ago during migrations to different hunting grounds. It would make sense for the body to become metabolically more efficient and for us to be driven to increased food-seeking behaviour during this period. It would have helped us to survive these times. Unfortunately, we still carry these metabolic responses to sleep deprivation and this leads to a rise in our weight set-point. Our evolved bodies are still hard-wired to react in exactly the same way to the voluntary sleep deprivation experienced during a TV binge as we do to the enforced sleep deprivation of a long and arduous journey – namely, by increased appetite, increased blood sugar, increased insulin resistance . . . and, finally, weight gain.

In the course of Step 2, we are going to try and reverse these changes, and use our understanding of how the body works to lose weight the smart way. We will try to sleep more and therefore decrease our set-point . . . and our weight.

First, we need to change our evening relaxation and sleep habits. But, remember, you are only likely to sustain this change of habit if it becomes more enjoyable and relaxing for you. One of the critical factors in becoming sleepy is to stimulate that secretion of melatonin – and the best way to do this is to reduce the ambient lighting in your home environment as the evening progresses.

We know that in areas of the world where there is no electric lighting, melatonin starts working at sunset, and then within two hours people will fall asleep (in rural African villages, sleep time is 9 p.m. and wake time is 5 a.m. as the first light starts to appear). To help you sleep you should try and get into a routine – just as you would if you were living closer to nature, when your pineal gland would automatically sense nightfall – and start to dim the lighting (and avoid screen time) two hours before bed.

In preparation for this step you need to have installed lower wattage bulbs, low table lamps and maybe even dimmer switches. You can make this more primeval – and therefore probably even more effective – by using unscented candles for some of your lighting (carefully!). An hour or two of low lighting and you will start to feel sleepy.

Sometimes an earlier sleep time than you've been used to can be difficult to adjust to. A relaxing hot bath, low music and some herbal tea can help. If you have lost the habit (or never had it in the first place) of reading in bed, take it up again. Books can take you into another world and reading can often have a soporific effect.

Even if you don't fall asleep immediately, don't worry about it, just enjoy resting your body and try and have happy thoughts and pleasant memories. Sleep will come and you will feel more refreshed, enabling you to enjoy the day ahead when you wake.

These new habits should become part of your normal routine. Obviously, there will be times when you are out and about until late, but try and slip back into and enjoy looking forward to your routine night in. In time an increase in your sleep to eight hours or so per day will help your mood, health and metabolism – and your set-point will drop. You will lose weight naturally.

SIXTEEN

Your Personal Blue Zone

Improving Cell and Muscle Metabolism

STEP 3 – PRIME YOUR CELLS

In his book *The Blue Zones* (2008), Dan Buettner identified five areas, or zones, in the world where people live the longest. He visited these areas and studied the lifestyle and habits of their populations to try and discover the secret to their health. He identified several factors that each of the zones had in common and concluded that these traits were responsible for the remarkable fitness of the populations living in them. The characteristics he identified included: a plant-based

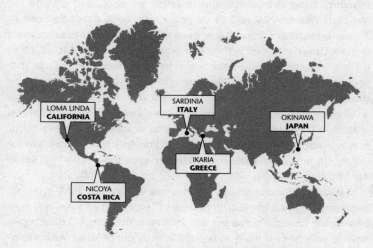

Figure 16.1 The five original blue zones

Source: Adapted from an idea in Dan Buettner (2008), *The Blue Zones*. National Geographic.

, moderate but not excessive physical activity, low stress mily-based communities with good social interaction.

The blue zones are located in Okinawa (Japan), Nicoya (Costa Rica), Sardinia (Italy), Ikaria (Greece) and Loma Linda (California). When you check these places out on the map, they have something else in common – they are all coastal areas and would have had an abundance of fish in their diet. If you look closer, these healthy communities had no problems with obesity. My guess would be that, in the presence of so much fish and vegetables in their diet, and the absence of man-made polyunsaturated vegetable oils, the omega-3 to omega-6 ratio of the populations would be normal (1:1 to 1:4), compared with the increased incidence of omega-6 in the Western diet, contributing to the health of the inhabitants. The cell walls of the people in these areas (every cell in every inhabitant that had lived there for some time) would have taken on the imprint of the food they were grow-ing, catching and consuming. With such a healthy ratio of these fatty acids on board, the incidence of inflammatory Western-type diseases would be minimized, and this would explain the longevity of the population. It would also explain the lack of obesity.

In Step 3 of this programme we will try and mimic the food envir-onment of the blue zones, and in doing so we should try and get our own cellular omega-3 to omega-6 ratio back down to normal. This will result in a lowering of the weight set-point as insulin starts working more efficiently (therefore you will need less of it) and lep-tin begins to be sensed by the brain, meaning that the message that you may be carrying too much weight will finally get through to the weight-control centre in your brain (the hypothalamus). This will result in an effortless increase in your metabolism and a natural reduction in your appetite – resulting in weight loss down to your new set-point. The other bonus is that you will have significant (or improved) protection against any Western-type inflammatory disease. Are you ready to get healthy and slim?

If you have followed Steps 1 and 2 you will already have given up sugar and highly refined carbohydrates (like wheat) and you will now be sleeping for eight hours per day. However, the omega-3 to omega-6 ratio in your body is likely to be significantly out of kilter, with a mas-sive excess of pro-inflammatory, obesity-stimulating omega-6. Your current ratio is probably in the region of 1:15 to 1:20, maybe even more. In this step, we will try and redress the balance by consuming a more natural ratio of the omega-containing foods. Once we do this, our bodies will, as we have previously stated, take on the imprint of the food in our present environment that we are consuming.

The 'Fat' Vitamins

As a reminder, there are two fats in nature that we are unable to make inside our body – that is why they are called essential fatty acids. These two fats have important metabolic and inflammatory functions in our cell walls. If we do not have them in our diet, we get very sick. They are the fat equivalent of vitamins.

We need to bear in mind when talking about the omega-3 to omega-6 ratio in our bodies that these two fatty acids, the flexible but easily oxidized omega-3 and the more rigid but more stable omega-6, *compete with each other* for space on each of our cell's walls. This is critical: if one of the omegas is taken into your body in excess, it will dilute the other omega fatty acid, even if your consumption of the other fatty acid seems adequate – and the composition of the cell wall will change.

Imagine that the cell walls in your body are like the walls in a room that you need to paint. You want to paint the wall a sky-blue colour to match the colour of an Ikea cabinet you have just made. To do this you have to bring into the house and mix together exactly the right proportion of blue and white paint – the colour depends on the ratio of the two paints to be used. If you bring too much blue paint into the house, you will have a wall that is too dark and doesn't match the colour you wanted. Think of the fatty acids omega-3 and omega-6 as being like different coloured paints on your cell walls. Too much of one or the other and the colour will be wrong. At this moment in time you have a dramatic excess of omega-6 in your cell walls – which will drown out the 'colour' of omega-3. The rigid and pro-inflammatory omega-6 fatty acid has transferred from the food that you have eaten onto your cell walls, drenching them in the wrong colour for optimal health.

This part of our dietary plan aims to rectify the balance. Just as in our paint analogy, we have to bring into our bodies the correct amounts of each of the omegas. We have to consume foods that will make our cells shine with just the right omega colour to stimulate our metabolism, improve our general health and lower our weight set-point even more.

Sunshine Food

It is relatively straightforward to recognize foods with high levels of omega-3 or with greater levels of omega-6. As we saw in chapter 9, omega-3 is found in chloroplasts, the cellular engines in plant leaves (and algae) that convert sunlight into biological energy. It's a message to our bodies that summer is here and food is plentiful. So any foods that contain green leaves will have great omega-3 levels. Any animal (or fish) that has consumed leaves (or algae) will also have high omega-3 levels.

Omega-6 foods – the ones that you should be aware of and cut down on – are the autumn foods that might cause a slower metabolism and pre-winter weight gain. Omega-6 is contained in nuts and seeds (which includes all grains).

We learned in chapter 9 that an excessive but artificial quantity of omega-6 fatty acids has entered our food supply in the form of vegetable oils. The emergence of these oils coincided with government advice to cut down on saturated fats – they are derived from easily grown seed crops that are not natural human foods (sunflower seeds, rapeseed etc.). They have to be refined in a similar way to crude oil in order to make them safe, yet they have been labelled healthy on the back of flimsy nutritional research, and they are now well and truly entrenched in our food supply.*[1]

It is not just in the large bottle of cooking oil in your kitchen, containing massive amounts of omega-6 which will soon be inside you, creating metabolic and inflammatory havoc. No, these types of oils infuse most Western processed foods – from margarine, fried foods (chips, doughnuts) and baked foods to snacks (such as crisps) and vegetable shortening. The beauty of omega-6 oil, from the food industry's point of view, is that it is relatively stable so it can be added to foods that need to be transported long distances and sit on the shelves of shops for months waiting to be sold. Omega-3 fatty acids, on the other hand, need to be taken out of these foods as they make the food go off too soon (affecting the food company's profits).

The reason for the massive disparity in the omega-3 to omega-6 ratio seen in Western populations is not, then, a sudden deficiency of omega-3-containing foods (like fish). The problem is the massive excess of omega-6 in our diets – in the form of vegetable oils and processed foods. Many health commentators suggest that we should increase our consumption of omega-3 foods in order to solve the

---

* There has been only one single study suggesting that vegetable oils lower the risk of heart disease: this study is often quoted on food labels – but scores of further studies have failed to find a protective effect.

problem, but this logic is flawed. If we are taking in such a massive load of omega-6, then increasing omega-3 slightly will have minimal effect – it will still be heavily diluted.

A Splash of Corn Oil

To get the amount of omega-6 in these oils into perspective, let's look at an average serving (2 tablespoons) of corn oil used for shallow frying. This contains 14,000mg (14 grams) of omega-6 and only 300mg of omega-3. If we take a generous serving (150 grams) of Atlantic salmon (one of the highest omega-3-containing foods available), we take in 3,000mg of omega-3 and minimal omega-6. Therefore, every time we fry food using corn oil, we would need to eat four large portions of omega-3-rich salmon to match the amount of omega-6 in the oil. Alternatively, we could take omega-3 capsules (usually 500mg per capsule) from the chemist – and take twenty-eight of them! You can begin to see why our omega-3 to omega-6 ratio is so high when we consume foods containing vegetable oils.

There is a simple way to even out your omega-3 to omega-6 ratio (simpler than force-feeding yourself fish or cod liver capsules) and that is to *replace your vegetable oil*. Natural cow's butter (saturated fat) or virgin olive oil (monounsaturated fat) have much less omega-6 than vegetable oils. So, to begin with, I would suggest frying and baking foods using these traditional alternatives.

'What about canola oil?' you may ask. The label on the bottle says it is high in omega-3. If you look at the quantity of omegas in a tablespoon of this oil (1,200mg of omega-3 and 2,600mg of omega-6) the ratio of 1:2 is not too bad. It certainly does contain a fair amount of omega-3. The problem arises when you start to cook with this oil. At the high temperatures required for frying, most of the omega-3 will be degraded and will not be useful – the omega-3 health claims on such bottles are therefore just a marketing ploy to persuade you to buy it.

Rule 1

Use butter and olive oil instead of vegetable oil to fry and bake with.

* Remove vegetable oil from your home
* Buy butter and olive oil.*

---

* The olive oil should be bought in a tin or, if in a glass bottle, it should be stored in a dark cupboard as sunlight degrades the healthy antioxidants in the oil.

Vegetable oils hide themselves, in large amounts, in many other foods that are common in the Western diet. Remember that the cannabinoids in omega-6-containing foods can also be slightly addictive, so you may have to use the psychological tricks we outlined in chapter 14 to help free yourself from these foods.

Rule 2

Do not eat foods that contain, or have been cooked using vegetable oil:

- fast foods
- crisps, fried snacks, health bars
- ready-made cooking sauces
- margarines and oily spreads.

Fast food from the high street is cooked in, and therefore contains large doses of, omega-6. Examples include:

- KFC chicken (13,500mg)
- Burger King onion rings (10,500mg)
- Burger King Double Whopper with cheese (10,300mg)
- Domino's Pizza (approx. 3,000mg/slice)
- fries (around 4,000mg)
- sauces, e.g. McDonald's Creamy Ranch Sauce (10,700mg per ½oz).

You might expect that a Subway tuna sandwich would be a reasonably healthy omega-3 to omega-6 choice. Unfortunately, this tops the list of omega-6-containing fast foods with 14,000mg per serving (vegetable oils are used to make up the mayonnaise in the tuna salad).

And it is not just the obvious fast-food high-street shops that you should be wary of. Many delivery or fast-dining restaurants, such as Indian and Chinese restaurants, use vast quantities of vegetable oils in their foods and sauces.

Snacks can also contain large amounts of omega-6 – such as potato crisps (8,900mg), tortilla chips (8,800mg) and granola bars (4,600mg). Microwave popcorn (22,000mg) and crisps manufactured from dried potatoes (18,000mg) are particularly toxic.

There are many ready-made cooking sauces now available in supermarkets. They make cooking a delicious meal easy – just fry the meat and add the sauce and you have a meal. However, you should also be wary of these as they contain copious quantities of vegetable oils – and therefore omega-6. This is why it is much better if you can learn to make food from wholesome ingredients. Sauces can be made at home using butter, milk, olive oil etc. OK, you can't store them on your shelf after you have made them for six months, but that is because they are real foods containing fresh healthy ingredients. Remember fresh food lasts for a matter of days; foods containing omega-6 last months. This is how you can tell the difference.

Rule 3
Avoid food containing very high omega-6:

Meats and meat substitutes
Other foods that you should be aware contain omega-6 in significant levels include cured meats and tofu. For example:
- chicken sausage (5,900mg in each)
- frankfurter (2,100mg)
- salami (3,600mg/100g)
- tofu (fried, 10,000mg per 100g).

Nuts
A special mention should be made of nuts and dried seeds, and foods made up predominantly of nuts and seeds, sold as 'healthy' snack bars. The omega-6 content of 50 grams (equivalent to a small packet) of the following nuts is:
- sunflower seeds (18,000mg)
- almonds (6,500mg)
- cashews (4,200mg)
- peanuts (roasted, 8,500mg).

Walnuts are regularly mentioned in the press as a great source of omega-3. They do indeed contain a lot – in fact 50 grams of walnuts contain around 4,500mg of healthy omega-3 (double the amount contained in a fillet of salmon). There is one catch though (usually overlooked in health articles). Walnuts also contain a massive 19,000mg of omega-6 (per 50 gram serving). So the large quantities of healthy omega-3 that walnuts contain are rendered useless by the omega-6.

Rule 4

Choose meat and fish with higher omega-3 levels:

- grass-fed beef (check the label carefully)
- lamb (usually 100 per cent grass-fed)
- line-caught fish (if they are fed in a fish farm, their diet will be grains and they will have higher omega-6 levels)
- canned fish (in brine, *not oil*: very healthy omega-3 profile)
- avoid grain-fed chicken.

Omega-3 comes from green-leaved plants (and green algae in the sea) and any animal or fish that eats leaves, grass or algae. Remember that many farms feed their livestock an unnatural grain-based diet, containing omega-6, in order to make them grow bigger faster (it works for animals as well as humans). The livestock from these farms will contain meat that has a poor level of omega-3 and a high level of omega-6 (just like us humans). Therefore, meat from these farms should be avoided if possible. Unfortunately, farms that feed their livestock grains are now the norm and you will have to source grass-fed animal meat carefully. Almost all chicken and pork and most beef are fed grains, so try and avoid these. Most lamb is still left to graze on grass, so this is a better option.

Fish farms are also very common. Just like land-based animals, salmon that has been fed grain will have a much poorer omega-3 to omega-6 profile than line-caught fish from the ocean.

Rule 5

Eat as many fresh vegetable and dairy products as you wish.

Vegetables and dairy products have low levels of omega fats and their ratios tend to be healthy, so eat these foods in abundance.

Summary

To optimize your omega-3 to omega-6 ratio, follow some simple rules. Eat lots of greens, and eat lots of meat and fish that have eaten their greens; you can also include dairy products (and, yes, butter is OK). Cut out vegetable oil, seeds (including grains) and processed foods. Because the good foods tend to be fresh for only a short period of time, you will have to shop regularly and you will have to cook.

Michael Pollan, in his great book *In Defence of Food*, suggests some simple-to-remember rules when out shopping.

- Don't buy anything that your great-grandmother would not have recognized as food
- Don't buy food that *doesn't* go off
- Don't buy foods that are packaged – especially if they contain more than five ingredients, or are labelled as healthy: 'low fat', 'no added sugar' or 'low cholesterol' are typical health claims to be wary of.

The Greengrocer, Butcher, Fishmonger Diet

My simple eating rule would be to try and buy all your food from the greengrocer's (a traditional one that just sells fruit and vegetables), butcher's (one that sells dairy products as well as fresh meat) and fishmonger's or fish counters. If most of the food you consume is made up of fresh vegetables, meat, fish and dairy products, and is home-cooked (without vegetable oil), you will be on the right track to improving your cellular metabolic health.

STEP 4 – TONE YOUR MUSCLES

Can regular exercise lower your set-point and therefore your weight?

Those who adhere to the energy in/energy out equation see exercise as just as important as calorie intake – they think it's all about counting. This is why the gym industry is such big business. However, we know that metabolic adaptation* can be more powerful than any gym membership, accounting for many hundreds of calories extra, or less, per day, every day. If your body doesn't want to shift body weight, it will adapt to exercise and go into energy-conservation mode. In addition, we know that if you exercise too heavily then it is likely that your body will direct you to refuel by increasing the production of the appetite hormones. This is the reason there is a juice bar or snack bar attached to most gyms.

---

* Metabolic adaptation is the decrease or increase in our metabolism in response to dieting or over-eating – as discussed in chapters 1 and 3 in Part One.

The Hunters vs the Office Workers

A famous study compared the amount of energy expended by hunter-gatherer tribes in Tanzania (the Hadza) to Western city dwellers (in New York and London), using the most accurate measure of energy expenditure, a technique called doubly labelled water.[2] We know that hunter-gatherers spent a large part of their day active – walking or running – whereas the average city dweller is sedentary – maybe having a short walk from the car or the station to the office. When scientists compared the total amount of energy burned over a thirty-day period, they found that there was no difference! The Hadza hunters and the city dwellers used up the same amount of energy despite one being active and the other being sedentary.* The researchers concluded that the Hadza tribesmen became hyper-efficient when resting, burning much less energy at night than the city dwellers. What the scientists failed to address was that the city dwellers were probably hyper-metabolizing throughout the day to compensate for over-eating. My hypothesis (as explained in chapter 3) is that further research would have revealed that city dwellers were hyper-metabolizing – burning their excess calories through an activated sympathetic nervous system (leading to high blood pressure) and through adaptive thermogenesis (leading to loss of energy to heat).

The message of the study was that exercise activity is easily compensated for – the body will fight to maintain its weight set-point whether that means metabolic hyper-efficiency in active people who don't over-eat, or metabolic inefficiency in those who are sedentary and over-eat: the crux is the set-point.

The Stressed, 'Hypertensive', Overweight New Yorker

The research study did confirm one expected difference between the two populations of people studied – their size. When comparing the slim Hadza to the fat New Yorker it is not a calculation of the total calories taken in compared to the total calories expended in exercise – this is far too simplistic. No, the New Yorker has been exposed to environmental signals that have raised their set-point. Yes, we know the restaurants in New York are some of the best in the world, but it is the quality of the food and not the total calories

---

* This took into account the relative sizes of the Hadza hunters and city dwellers. In actual fact, the hunter-gatherers used up less energy than the city dwellers because they were smaller.

consumed that determines the set-point. The New Yorker would have been exposed to large amounts of processed and fast foods – leading to a major derangement of the omega-3 to omega-6 cell membrane ratio, as well as to chronically high insulin requirements. Add to this the stresses of city life, and maybe the disturbance in melatonin levels due to poor day/night differentiation, and you have multiple factors elevating the set-point higher and higher.

How Does Exercise Work?

So, we have a dilemma. The active Hadza tribe didn't expend more energy than the sedentary New Yorkers. If regular exercise is compensated for by metabolic adaptation, if it makes us more efficient at resting and we therefore don't burn off more calories, how does it work?

I think that we can safely say that exercise *does* produce weight loss (otherwise gyms would not be so popular). It's just that the weight is not lost by the simple energy in/energy out equation that most people imagine. It is lost because exercise itself causes our weight set-point to decrease. Only once this has happened will the body let that energy go (probably in the gym) and the weight will come down.

Regular strenuous exercise will cause two major changes that affect our weight set-point:

1. Decreased cortisol (the stress hormone)
2. Improved insulin sensitivity (leading to lower insulin levels).

We can summarize as follows:

Exercise → lower cortisol → lower set-point
Exercise → improved insulin sensitivity → lower
insulin → lower set-point

The lower set-point then directs the body to lose weight.

The importance of exercise is not therefore about the calories expended, it is about how the activity improves your metabolic health by decreasing your cortisol levels and improving insulin sensitivity. Exercise can have a bigger impact on weight if you can find the time in your schedule to train like an athlete. If you perform intense activity for two hours a day, expending 1,000kcal/day, most days a week, then this is something that metabolically the body cannot ignore. But for most people this is not something they can fit into their daily routine.

One of the other major benefits of exercise is that it produces a significant increase in your 'good cholesterol', HDL. This trumps all other types of bad cholesterol and will significantly reduce your risk of heart disease.

Activity Rules

- Choose an activity that you enjoy
- Choose an activity that is practical for you
- Exercise two or three times per week for a minimum of twenty minutes
- Exercise enough to sweat (this is important for it to work)
- Avoid endurance exercise.

It's important for the activity to be one you really enjoy – not everyone likes going to the gym. Personally I am slightly intimidated by the muscular guys sashaying naked around the changing room, sniggering at my (not so muscular) body, and I imagine that many other people also feel uncomfortable in that environment. Find an activity you will look forward to and really enjoy, one that can enrich your life. Perhaps swimming, yoga, tennis or squash? If time is an issue, and you used to like cycling, then cycle to work and back occasionally. If you prefer team sports, there are football, netball or hockey. You might just like to walk, and a brisk walk with a bit of an incline can be good and enjoyable outdoor exercise; or, like me, you might want to go for a jog (or even a walk and a jog). If you are not very good at sport, then look out for lessons or a club and take up something new. You may choose to get a rowing machine or treadmill in your own home and watch TV or listen to music when you exercise. The main point is it has to be enjoyable for you; otherwise you will stop doing it after a while. Also remember, exercise isn't about counting calories off – we should be exercising to improve our metabolic health, to lower our insulin and cortisol levels and also to improve muscle tone. Muscle health, as we are about to learn, is essential for our weight regulation.

Keep Your Muscles Strong
In addition to improving your metabolic health, exercise contributes to muscle health. There are some societies in the world, particularly the

Middle East, where extreme sedentary behaviour, especially amongst women, is almost expected – it is the cultural norm. When you are not expected to perform any household chores, and it is unladylike to walk long distances (unless in the shopping mall), then over time your muscle mass will shrink and you will develop a condition called sarcopenia (meaning small, withered muscles). As we found out in chapter 3, by means of metabolic adaptation in our muscles, we literally burn off the excess calories consumed (this is known as thermogenesis). Therefore, when muscle mass shrinks, our ability to burn off excess calories is compromised. If you combine small muscle mass with larger quantities of calories (for instance, by eating sugary snacks), only one thing will follow: significant and rapid weight gain. This is why Middle Eastern women now have staggering obesity rates nearing 50 per cent.

The key message for you to keep in mind is: maintain muscle strength and muscle mass. Keep your essential muscle organ healthy. If you are a 'couch potato' in the evening, don't worry, as long as you do some short, intense, muscular activity before your down-time. If you don't have time for, or don't like, the gym, try something like the 7-Minute Workout app. It will ensure all-round muscle health – and allow you some me-time.

STEP 5 – REDUCE YOUR INSULIN

We have now arrived at the final step in optimizing your weight. If you have made it to this point, congratulations! I hope that you are enjoying your new, much healthier life. So far you have given up sugar and highly refined carbohydrates, improved your sleep, optimized your cellular health and are performing an exercise that you enjoy. In the final part of the programme we aim to reduce your carbohydrate intake slightly more, and by doing so reduce your insulin requirement – as we know by now, this will reduce your weight set-point.

There are different varieties of low-carb diets from the extremes of ketogenic dieting (discussed in chapter 12) to the more moderate 'low GI diet'. We are going to go for something in between – something that will be effective, but also can be sustained as part of a normal eating routine.

Usain Bolt vs Mo Farah

You have probably heard the term *glycaemic index* (GI). It is used to describe the speed at which a food releases its carbohydrate energy into the bloodstream. The higher the GI of a food, the faster it releases its sugar energy. It is used to identify foods that will give you that sudden insulin spike I described in chapter 10 and is the basis for the low-GI diet. In this diet, participants are asked to avoid foods that have a high glycaemic index and to choose only foods that release their glucose slowly. Because the diet normalizes blood sugar fluctuations, it is particularly useful for diabetics on insulin. The type of GI foods that you are encouraged to eat while on the low-GI diet would include grapefruit (GI of 25), cherries (22), apples (28) and sweet potatoes (40).

The types of foods that you are told to avoid in the low GI diet are white potatoes (85), white bread (70), watermelon (72) and carrots (47).

But the glycaemic index does not tell the whole story about the food that you are eating.

Consider this question. Who would win in a race between Usain Bolt and Mo Farah? You can imagine them lining up against each other under the floodlights in the Olympic stadium in London. The starter is ready to fire his pistol. Many people would probably not think the question through and would automatically say, 'Usain, of course'. But they are assuming the race is over a short distance. They are answering the question: 'Who is the fastest?' But if after the starting pistol has been fired, and the lightning-fast Bolt sprints into an extremely healthy early lead, the finish line does not appear . . . then the runner with more stamina will have a chance. As the muscular bulk of Usain starts to cramp up, you can imagine Mo Farah gliding past him – and the MoBot will win.

The Glycaemic Load
In just the same way that runners should not be defined only by their speed, so a food should not be defined only by its glycaemic index. It is not just the speed at which the food releases glucose into the bloodstream that is important, but the total amount of glucose that is released. This is where I think the *glycaemic load* becomes much more important – in predicting total insulin levels – than the glycaemic index.

The glycaemic load defines the full effect that a portion of food will have on your blood glucose level (not just the speed). One unit of glycaemic load has the same effect as consuming one gram of glucose (4kcal). It is influenced by the portion of the food that is eaten, so doubling a portion of food will double its glycaemic load.

For example, if we adhere to the low-glycaemic-index diet principle we will see that watermelon is a fast releaser of glucose. It just sprints into your bloodstream with a GI time of 72, whereas a low-fat yogurt is much slower (GI of 33). But the food equivalent of the MoBot has much more energy stored within it. The yogurt's glycaemic load (GL) is 16 per pot, compared with a slice of watermelon that has a GL of only 8, half the amount. So in the long run a pot of yogurt increases insulin by double the amount compared to a cup of watermelon, despite watermelon having a much higher glycaemic index.

Here are some examples of the glycaemic loads of common foods:

Staple Carbs		Fruit and Vegetables		Meat and Dairy	
White potato	29	Orange	4	Beef	0
Sweet potato	20	Apple	5	Chicken	0
White rice	24	Banana	10	Eggs	0
Wild rice	16	Grapes	9	Milk	9
White bread	16	Green beans	4	Cheese	1
Brown bread	10	Tomato	2	Pulses	
Rice pasta	21	Spinach	2	Beans	12
Rice noodles	21	Carrots	2	Chickpeas	20

Table 16.1 The glycaemic load of common foods

Note: Portion sizes: potato (large size), rice, pasta, noodles (150g serving), bread (2 slices), fruit (1 item), grapes (1 handful), vegetables (1 medium cupful), milk (250ml), cheese (half a cup, diced), pulses (half a can, 200g).
Source: USDA National Nutrient Database for Standard Reference, April 2018.

There is also an extensive table of the glycaemic loads of different foods in Appendix 2. You will see that meat, fish, eggs and cheese have a GL of zero and that most fruits and vegetables have a low GL. Our blood glucose mainly comes from our staple carbohydrate foods (potatoes, pasta, rice and noodles). I do not want you to give these staple foods up and start to crave them, or even to develop the unpleasant side effects of the ketogenic diet – but I think that it will be beneficial for your insulin profile and therefore your weight set-point (and ultimately your weight) if you slowly try to reduce the overall amount of glucose entering your bloodstream every day. By reducing your total glycaemic load you will be able to do this. As I have explained throughout this book, it is not the total number of calories that you consume that matters, it is the *quality* of the food that is important. When you cut down on your portion size of staple carbohydrates, you should compensate by topping up on low GL vegetables and high-protein and high-fat foods.

Measuring Your Daily Glycaemic Load

Before you start trying to reduce your glycaemic load, you should measure your current level. You can calculate the amount of carbohydrates that you consume every day by using an app on your smartphone such as MyFitnessPal. You may need to invest in kitchen scales (if you haven't already) to get an idea of the size of your portions by weight. The app can then calculate the GL of each individual food and add up the daily total.

150, 100, 80 or 60 Grams?

Most people who are not dieting will consume in excess of 300 grams of carbohydrates per day: that is a total glycaemic load of over 300. I think that a good starting target for your glycaemic load should be 150g per day. This should be easily achievable, especially as you are already avoiding any significant carbs for your breakfast. Once you become more aware of the high-carb foods in your diet, the next step should be to get your daily GL down to 100. Do not rush this. It is much better to make slow planned changes over weeks rather than days.

The ultimate target could be as low as 80, but this will depend on how your body is responding to the changes and how you are feeling, whether you are able to cope easily with these changes and are enjoying their health benefits. Remember, if any part of the programme is not enjoyable it is much less likely that it will become part of your daily routine, and therefore part of you.

Don't Go Keto

The aim of our plan is to lower your insulin levels by reducing your total carbohydrate intake. But we do not want your carb intake to go so low that your liver runs out of reserves and you become ketogenic. Sometimes this can happen if you are simultaneously exercising and cutting your daily carb intake. If you feel particularly weak, or experience symptoms of keto flu such as headache, nausea or vomiting, then it might be that you have exhausted your liver's reserves of carbohydrate (your 'battery' is flat). You should be aware that exercise can drain your liver of carbs and tip you into ketosis if you are not replacing them.

So here is the good news: you need to replace those carbs that you burn during exercise – you can add them to your daily allowance. Most moderate exercise, such as jogging, working out in the gym or playing football or tennis, will burn 250–350kcal per half-hour session. This energy is taken from your liver: 300kcal translates to 75 grams of carbs, which would be the equivalent of an extra-large baked potato, a portion of rice *and* a banana – all on top of your daily GL target. Seems OK, maybe worth the effort . . . and certainly better than topping up with a single Snickers bar (270kcal).

Finishing Line

I hope that you have enjoyed the steps of this programme and that you are finally winning the diet war – once and for all. The changes to the working of your body that each of these steps makes can be variable, and can take weeks, even months. But if you persist with the way of eating and living set out in this book those changes will soon become ingrained into your body. Eventually your body will reflect the way you live. Your weight set-point will be permanently lower, meaning easy, seamless weight regulation and a long-term improvement to your metabolic health.

EPILOGUE

Why Do We Eat Too Much?

When my very helpful editor at Penguin suggested we name this book *Why We Eat (Too Much)*, I had to sit down and think it through. A more appropriate name for the book would have been something like 'Why Do Some of Us Store Too Much Energy (and Others Don't)', but this was clearly too long-winded and wasn't going to catch the eye of potential readers, or get the messages of the book out there into the public domain (the main reason for writing it in the first place).

Before reading these pages, many people attempting to answer this question would have replied, 'Because we are greedy', or 'Because food just tastes too good these days'. But as we have learned in this book, it's a lot more complicated than that.

In truth, as we saw in chapter 1, most of us eat too much nowadays, a lot more (500kcal/day) than we did thirty years ago. But we also learned that despite eating more we also metabolize more. We are able to adapt to over-eating and burn off most of the excess energy effortlessly. Therefore weight gain doesn't happen as dramatically as we would expect. Remember the analogy with a delivery of logs for your fire in chapter 1? If more are delivered, you tend to burn more. If you eat more, you metabolize more. But that doesn't explain why many of us seem to be storing some of those extra logs for another day.

The Three Houses

Imagine three identical houses next to each other in the countryside. Each house is heated by a log fire and each house has a storage shed for logs outside the front door. All the houses receive an ample delivery of wood every day, more than enough to keep warm.

The first house has minimal wood in its storage shed and yet there is a constant plume of smoke from its chimney, and some of the windows are open to let heat out.

The second house's storage shed is almost full of logs. Its plume of smoke is less than house one's, and its windows are closed. The owner is clearly being frugal with his wood until he has more logs delivered and the shed is full.

In the third house the wood shed is damaged. This happened when it was over-filled. But there is a mountain of wood piled up by the side of the house. Despite this, the owner appears to have doubled his delivery order.

So three identical houses, but why are there three very different stores of logs?

The first house sits next to a forest. Its owner knows that there will never be a shortage of wood. There is no need to store much and in fact excess logs are regularly burned.

The owner of the second house is more cautious. His delivery man went on strike last year, leaving him with dwindling stocks of logs and a cold house. In addition, he recently heard on the radio that an icy spell of weather is approaching. Understandably he wants his shed to be full to the brim with logs.

In the third house, when the wood shed got damaged the delivery man from the timber merchants helpfully cleared most of the logs and stacked them at the side of the house – out of view of the owner. The owner didn't know this and, concerned that his supplies are running low, orders extra wood to be delivered. Unfortunately, most of his extra order is ending up stacked against the growing log-mountain out of view.

The owners of the three houses have very different interpretations of the outside world, and three very different stores of wood. But note that there was no difference in the amount of wood that the first and second houses ordered. House one burned that wood, house two was more frugal and stored it for an icy day. The only house ordering excessive wood was the third one, and this was only because its owner thought that his wood shed was almost empty. He wasn't aware of his massive stock of logs because they were out of view.

It's helpful to remember this analogy when thinking about obesity, the currently misunderstood disease affecting a quarter to a third of our population. Just substitute our hypothalamus (our weight-control centre in the brain) for the houses' owners, food for the delivery of logs and fat for the store of wood. Switch the delivery man's strike for a diet and the icy spell for Western food signals. Switch the three houses for people: the first house is a naturally slim person, the second house an overweight person and the third house is a person with leptin resistance (their log store is excessive, but invisible to the owner), causing fully fledged obesity.

Our old-fashioned understanding of obesity is slowly being challenged. Many scientists are realizing that it is not the *quantity* of calories that are available in the food supply to a population that will affect obesity levels: natural foods do not make populations fat. No, it is the *quality* of the food available that causes obesity. If a grain-, oil- and sugar-based diet is fed to any kind of population, whether it be a herd of cows, mice living in a lab or a continent of humans, the same effect will occur: high levels of obesity.

The traditional understanding of obesity as a lifestyle choice is of great value to many lucrative and powerful industries. The diet industry, the gym industry, the food industry and the pharmaceutical industry all have a vested interest in perpetuating this view. As we've explored, the food industry makes the processed foods that cause obesity. Without obesity, the gym and diet industries would not exist (they didn't 100 years ago), and without obesity many of the highly profitable drugs that the pharma industry sell would not be needed.

If the mainstream view of obesity was found to be flawed, people would soon realize that Western food products are slowly poisoning them. They would switch to natural foods – no calorie-counting required. But such a huge change is unlikely. It's hard to imagine a tax on processed foods coming soon, one that would fund a public awareness campaign on healthy eating – rather than the half-baked calorie-counting campaign we see now.* What is needed is a large-scale, professional media assault on the population's psyche that encourages people to eat natural instead of processed foods, to cook and embrace food culture. This could work – this could be the solution – but not just yet . . .

But instead of waiting for this change, you could try and change

---

* The amount spent annually on advertising products in the food industry is *one hundred times more* than is spent by the government on healthy eating campaigns.

your own life, and your weight, by following the steps in this book: *not* guaranteed to result in a 10kg (22lb) weight loss in ten weeks, but maybe 20kg (over 3 stone) or even 30kg (4 stone and 10lb) . . . over two or three years? And certainly guaranteed to lead to improved long-term health and hopefully happiness. And, as an ultimate bonus, you will never have to buy another new miracle diet book ever again!

APPENDIX 1

The Cholesterol Debate

In the Acknowledgements pages of this book, I could have added a statement: 'This book would not have been possible without cholesterol.' But the cholesterol debate is so fundamentally important that I decided to give it its own section, in order to clarify some scientific points for those readers who are interested.

If the 1960s diet–heart hypothesis, which linked cholesterol with heart disease (see chapter 8), had not existed, if scientists had not convinced governments of their theory, and if the governments had not advised their people to ditch saturated fat – then I don't think that we would have needed this book. The diet–heart hypothesis, which was supposed to halt the rise in heart disease, led to a chain of events culminating in another public health crisis – obesity.

We have all been subjected to public health campaigns, and media articles, explaining how saturated fat leads to heart disease – they have been omnipresent for fifty years. Once a critical mass of people, thought to be 10–25 per cent of the population, become true believers in an idea, then the rest of the population will adopt it.[1] This is what happened with the diet–heart hypothesis, and this is why our populations now have a fat phobia.

Most people living in the West (including doctors) will visualize the link between cholesterol and heart disease this way: if you eat food containing saturated fats (like red meat), this will lead to high levels of cholesterol globules in the blood which can somehow clog the blood vessels and cause narrowing of the coronary arteries, and you will risk heart problems. This picture is absolutely ingrained into our society's psyche. Such thinking is an integral part of everyday life and normal conversations about health. Now, when you see a fatty steak or sausages (lots of fatty offal), you tend to recall this picture – with the greasy fat in the food clogging your blood vessels. We are now very wary of steak, eggs, cheese and whole milk (unless we are French). As the cholesterol message continues to be reinforced (because the number of believers is now so high), red meat and all dairy products – natural foods that have made up a good proportion of our diet for thousands of years – are seen as bad for us.

Our governments advised us that it would be better for our health if we switched to a diet with less saturated fat and replaced it with grains (plant seeds) and vegetable oil (plant seeds). Food companies also followed the government advice, but had to add more refined wheat and sugar to make the low-fat processed foods more palatable and therefore commercially viable. Our new diet, high in refined carbohydrates, meant that we needed to develop a snacking culture to help us cope with the blood sugar fluctuations between meals.

These alterations to our diet – the addition of high levels of omega-6 from vegetable oils, and the elevated insulin from sugar and snacking – led to the metabolic changes in our cells (insulin and leptin resistance) that encourage weight gain (chapters 9 and 10). These changes would not have occurred without the diet–heart hypothesis and the demonization of saturated fats.

The idea that saturated fat causes heart disease is now just as ingrained in us as the knowledge that smoking causes lung cancer. However, unlike the irrefutable science linking smoking and cancer, the diet–heart hypothesis was based on evidence that has subsequently been discredited. Ancel Keys' original study, linking the saturated-fat intake of a population to their rates of heart disease, was biased by the selection of only the countries that fitted the hypothesis (for example, France and Germany, which consumed high levels of fat but didn't have high rates of heart disease, were excluded).[2] Confounding factors – such as the knowledge that sugar intake was also high in countries with high saturated-fat intake – were ignored. Most recently it has come to light that scientists were paid large sums of money by the sugar industry to deflect the dietary blame for heart disease away from sugar and onto fats.[3] Accordingly the influential review published by these scientists shifted the emphasis of their findings for the diet–heart hypothesis to be accepted universally as fact.

So the reliability of this hypothesis has started to be questioned. There is increasing evidence that saturated fats from fresh foods (like red meat and dairy products) are *not* strongly associated with heart disease.[4] Unfortunately this message, from recent research, hasn't yet got through to policy-makers. As we have learned in this book, top researchers and scientists, and influential doctors, have vested interests. If an important public health message that they have been promoting for years is disproved, then their reputations will be discredited and the funding for their labs may dry up. This is the reason there is so much inertia about changing public health advice: too many people have their reputations and livelihoods tied up in the diet–heart hypothesis.

Let's look at up-to-date evidence to unpick the diet–heart hypothesis and see where we currently stand. Is it OK to eat saturated fat or not?

When the diet–heart hypothesis was gaining traction, the only relevant blood test available was to measure *total* cholesterol levels. We now know that it is not the total amount of cholesterol in the blood that is important in calculating the risk of heart disease, but the vehicle that carries the cholesterol within the blood. Cholesterol is a fat and so cannot dissolve in blood (think of balsamic vinegar and olive oil – they don't mix). When travelling in the blood, it needs to find its way inside water-friendly carriages. These carriages are called LDL and HDL (meaning low density lipoprotein and high density lipoprotein). The LDL can be either type A (small and tight particles) or type B (large and fluffy).

The Morning Commute

Think of cholesterol molecules, travelling in the blood, as being similar to people trying to get from home to work. Let's imagine that for their daily commute they must choose a vehicle to travel in. Some people take a big, hollow red bus (driven safely, by a trained bus driver) and others take tightly packed mini-vans (driven recklessly by freelance drivers). Think of the risk of heart disease as being analogous to the risk of traffic accidents. Immediately you can see that if everyone travelled in safe red buses there would be very few accidents, but when more and more people take mini-vans the accident rate goes up. It is not particularly the *number* of people travelling that affects the number of traffic accidents, but the *type* of transport that they choose. The risk for heart disease works in the same way: it is not the *total* amount of cholesterol travelling in the blood that is important, but the *type* of transportation. If cholesterol takes more LDL type B transport (red buses), the risk of heart disease is not increased; but if cholesterol travels using LDL type A transport (reckless mini-vans) the risk of heart disease does increase. The *total* amount of cholesterol in the blood is only significant in people who have hereditary high cholesterol levels. This condition affects 1 in 500 people and leads to heart disease very early in life (in their thirties or forties). It was this hereditary condition that tricked researchers into thinking that cholesterol levels in everyone were an important factor in heart disease risk.

I want to introduce our third vehicle now. Interspersed with the buses and mini-vans on our commuters' roads are police patrol cars. As we know, as soon as a police car is around, even the most reckless driver will behave for a while. In our analogy, the police patrols represent the effect of HDL on our cardiac risk. The more police cars on the road, the less the chance of accidents occurring: the more HDL in the blood, the less the risk of heart disease. Police patrol car numbers are *the* most important variable affecting accident rate: when their numbers drop, accidents rise dramatically. In the same way, healthy HDL levels confer much more protection against heart disease than any other factor.

The next question should be: what determines the type of transport that cholesterol uses? If the original diet–heart hypothesis was correct – that saturated fats cause heart disease – then we could conclude that increased consumption of these fats causes cholesterol to use LDL type A cholesterol (mini-vans) as their preferred mode of transport. But when the hypothesis was originally formulated, the type of vehicle that cholesterol travelled in was not known – only the total amount of blood cholesterol could be measured. We know from these early studies that a high dietary cholesterol intake does indeed increase total blood cholesterol slightly (the number of morning commuters in our analogy) and therefore that more cholesterol-carrying vehicles will be needed. But here is the catch – the higher-cholesterol traffic from saturated fat does *not* increase the ratio of LDL type A vehicles (mini-vans) compared to LDL type B vehicles (buses). The total number of LDL (buses and mini-vans combined) increases, but the proportion of good type B (buses) is increased and bad type A (mini-vans) is actually *decreased*. After eating saturated fat, the level of 'good' HDL cholesterol (police patrol cars) also increases, protecting against heart disease. This evidence would suggest that consuming saturated fats *does not* cause heart disease – and that the diet–heart hypothesis is wrong.

What other factors could change the cholesterol traffic in our blood-stream? Let's expand our analogy further. Let's say commuters have to walk to a bus stop to catch a bus, but that the crowded mini-van will come to their doorstep. If there is a torrential rainstorm, commuters are less likely to risk being soaked and therefore mini-bus traffic increases, leading to more accidents. In dietary terms, the rainstorm is produced by a different type of fat to cholesterol – known as trans-fats. As we have learned (see chapter 8), trans-fats are present in many processed foods, including cakes, biscuits and processed meats, and are also produced when vegetable oils are heated to a high temperature. Some of the earlier studies linking saturated fats to heart disease failed to take into account the effect of trans-fats on cholesterol commuter traffic, reinforcing the belief that saturated fats were dangerous.[5]

What about a snowstorm? Again, commuters will take the more convenient mini-vans rather than risk slipping over on the walk to the bus stop. The roads will be treacherous and again accidents will increase. For our cholesterol particles, the dietary equivalent to a snowstorm occurs when we take in – you guessed it – sugar.[6]

How about if the sun comes out (if you are reading this in a sunny country, bear in mind that a sunny day is a rare event in the UK)? Commuters want to enjoy their walk to the bus stop and avoid those sweaty, crowded mini-vans. In addition, there are more police cars cruising the streets (as they are less likely to call in sick on a sunny day). The result? Safe travel and no accidents. We can reproduce these idyllic cholesterol travel conditions within our bloodstream by something that costs nothing – exercise.[7]

So, to sum up, there are various dietary and lifestyle factors that influence the mode of transport that cholesterol takes and therefore affect the risk of heart disease. The most dangerous factors are sugar and the trans-fats in processed foods (snow and stormy conditions in our analogy). On the other hand, recent studies have suggested that saturated fats from natural foods are *not* a significant risk and, as we already know, exercise (our sunny day) is cardio-protective.

Good Cop . . .

Over the last ten years the growing acceptance that total cholesterol is not a dependable risk factor for heart disease has led to new additions to our everyday vocabulary: good cholesterol and bad cholesterol. 'Good cholesterol' is HDL (police patrol cars), while 'bad cholesterol' is still used to describe *both* types of LDL – type A and type B. This means that both our dangerous mini-vans and safe red buses are lumped together and described as bad. This has confounded the analysis of dietary risk, particularly of saturated fats, and muddied the research waters. It is almost as if some scientists are searching for vehicles through a thick fog. Why is this thick fog obscuring the truth about something so important to public health? Personally I am not sure, but suspect that vested interests of top scientific research institutions may play a part. We know that the direction of research is still, unfortunately, under the influence of the companies providing the funding for research labs. Scientists now must disclose their funding, but this still does not influence the direction of research that they take – it just makes it easier to find out whether there may be bias involved.

The biggest selling class of drugs in the world are the statins. According to a recent Informational Medical Statistics (IMS) report, revenues from cholesterol-lowering medications, including statins, reached $35 billion in 2010. These drugs have been shown to reduce total cholesterol levels in the blood by blocking some of its production in the liver. As well as reducing total cholesterol levels, statins also decrease the risk of heart problems in some patients. But many researchers now doubt that the effect of statins on heart disease is related to cholesterol: there is increasing evidence that they work to reduce inflammation in the cardiac blood vessels. If this is the case, then why does the American Heart Association (AHA), a body of experts that the rest of the world looks to for guidance, still support the diet–heart hypothesis and insist that LDL cholesterol (both types) is the most important marker of cardiac risk? And insist on a diet low in saturated fats? In fact, their recent guidelines have recommended lowering the threshold of blood cholesterol level appropriate for statin treatment,[8] guidelines based on a meta-analysis (summary of all previous studies) that excluded the important body of research on LDL subtypes.[9] It is as if this research does not exist and suggests a degree of bias. Many doctors around the world look to these guidelines in order to decide whether to prescribe statins, and if the diet–heart hypothesis remains valid then statins will remain bestsellers.

A particular saturated fat that has raised concern is *palmitic acid*. A report by the World Health Organization[10] has suggested that there is convincing evidence that the consumption of this type of fat can lead to heart disease. Palmitic acid is present in all types of meat and also in dairy products – but in small quantities. The pure form of palmitic acid is made by simply heating up palm oil to an extremely high temperature – this can be done on an open fire and it constitutes the main cooking oil that is used in African villages. Palm oil, when added to food, gives a nice texture and taste, and comes at a cheap price. As a result, large amounts are found in processed foods. It is, I believe, from these foods that the association of palmitic acid with heart disease arises – and not from the small quantities present in natural fats such as red meat, cheese and milk.

A recent independent meta-analysis of *all* the previous studies that looked at the relationship between saturated fats in the diet and the risk of death failed to show any increased risk, in particular no increased risk of developing heart disease, having a stroke or developing diabetes.[11]

Statins undoubtedly work in some cases, but I suspect that the rationale for prescribing them – high LDL and only slightly elevated total cholesterol levels – means that they are over-prescribed. The research certainly suits the pharmaceutical companies which produce the statins, but why are some medical associations (like the AHA) ignoring valid scientific research? What is the benefit for them? I will leave you to decide. Unfortunately, the perpetuation of the diet–heart hypothesis in the face of contradictory research means that dietary guidelines are stuck with the recommendation to avoid saturated fats from natural foods and to replace them with grains and artificial oils. Without these issues being addressed, our populations will still be guided to consume an obesogenic diet – and obesity will remain a major public health problem.

APPENDIX 2

Glycaemic Load and Omega-3 to Omega-6 Ratio of Common Foods

FOOD	SERVING SIZE	Cooked	WEIGHT (g)	GLYCAEMIC load	OMEGA-6 (mg)	OMEGA-3 (mg)
Greengrocer						
Potato, white	1 large	Baked	300	29	129	39
Potato, white, peeled	1 large	Boiled	300	26	96	30
Potato, mashed, with whole milk	1 cup	Boiled	210	16	81	35
Sweet potato	2 medium	Baked	300	20	103	6
Potato, oven chips	10 chips	Baked	133	13	232	21
Yam	1 cup	Boiled	145	13	43	8
Artichoke, French	1 cup	Boiled	168	5	264	100
Carrots	1 cup	Boiled	78	2	67	1
Broccoli	1 large stalk	Boiled	280	8	143	333
Spinach	1 cup	Boiled	180	2	30	166
Cauliflower	1 cup	Boiled	120	2	31	104
Cabbage, Pak choi	1 cup, shredded	Boiled	170	1	52	70
Cabbage, Savoy	1 cup, shredded	Boiled	145	3	26	33
Brussels sprouts	1 cup	Boiled	155	5	91	200
Asparagus	1 cup	Boiled	180	1	18	315
Green beans	1 cup	Boiled	125	4	70	111
Peas	1 cup	Boiled	160	9	30	131
Celery	1 large stalk	Raw	64	1	50	0
Tomatoes	1 cup	Raw	150	2	119	5

Source: Data courtesy of USDA National Nutrient Database for Standard Reference Nutrition Data; https://nutritiondata.self.com.

FOOD	SERVING SIZE	Cooked	WEIGHT (g)	GLYCAEMIC load	OMEGA-6 (mg)	OMEGA-3 (mg)
Tomatoes, canned	½ can	Boiled	200	6	108	5
Cucumber	½ cup	Raw	52	1	14	2
Beetroot	2 medium	Boiled	100	4	58	5
Mushrooms, portabello	1 cup	Grilled	121	3	242	0
Orange	1 fruit	Raw	140	4	43	15
Orange juice	1 cup	Juice	250	9	124	34
Apple	1 fruit, medium	Raw	180	5	78	16
Apple juice	1 cup	Juice	250	6	82	17
Pear	1 fruit	Raw	120	2	66	1
Banana	1 fruit, medium	Raw	120	10	54	31
Grapes	1 cup	Raw	150	9	55	16
Pineapple	1 cup	Canned	181	8	41	30
Butcher						
Beef, grain-fed, minced	1 steak	Raw	200	0	600	40
Beef, grass-fed, minced	1 steak	Raw	200	0	171	44
Beef burger	1 patty	Pan-broiled	82	0	270	56
Beef, roasted	1 portion	Roasted	200	0	660	240
Chicken, roasted	1 portion	Roasted	200	0	1,380	140
Lamb	1 piece	Roasted	230	0	1,631	1,095
Pork, ham	1 portion	Roasted	200	0	1,800	290

Source: Data courtesy of USDA National Nutrient Database for Standard Reference Nutrition Data; https://nutritiondata.self.com.

FOOD	SERVING SIZE	Cooked	WEIGHT (g)	GLYCAEMIC load	OMEGA-6 (mg)	OMEGA-3 (mg)
Dairy						
Butter	1 teaspoon		14	0	382	44
Margarine	1 teaspoon		14	0	4,357	42
Cheese, cheddar	½ cup diced		76	1	381	241
Cheese, cheddar imitation	½ cup diced		112	7	295	162
Cheese, brie	½ cup diced		72	1	369	225
Cheese, camembert	½ cup diced		123	1	553	336
Milk						
Full fat, 3.25%	1 cup		250	9	300	200
Semi-skimmed, 2% fat	1 cup		250	9	111	71
Skimmed	1 cup		250	9	12	5
Yogurt, low-fat	1 container		125	12	12	7
Yogurt, plain	1 container		113	4	73	30
Eggs						
Eggs, battery-reared	1 large	Poached	50	0	572	37
Egg yolk	1 large	Raw	17	0	600	38
Egg white	1 large	Raw	33	0	0	0
Eggs, fed flaxseed (omega-3 eggs)[1]	1 large	Raw	50	0	948	224
Eggs, fed fish oil[2]	1 large	Raw	50	0	624	229
Oils						
Sunflower oil	1 teaspoon		14	0	3,905	5
Olive oil	1 teaspoon		14	0	1,318	140

Source: Data courtesy of USDA National Nutrient Database for Standard Reference Nutrition Data; https://nutritiondata.self.com.

FOOD	SERVING SIZE	Cooked	WEIGHT (g)	GLYCAEMIC load	OMEGA-6 (mg)	OMEGA-3 (mg)
Canola oil	1 teaspoon		14	0	3,217	812
Sesame oil	1 teaspoon		14	0	5,576	40
Cod liver oil	1 teaspoon		14	0	126	2,664
Soyabean oil	1 teaspoon		14	0	6,807	917
Lard	1 teaspoon		14	0	1,428	140
Palm oil	1 teaspoon		14	0	1,228	27
Fishmonger/Fish Counter						
Cod	1 fillet	Baked	180	0	10	310
Haddock	1 fillet	Baked	150	0	18	400
Salmon, wild (line-caught)	½ fillet	Raw	200	0	408	3,000
Salmon, farmed	½ fillet	Raw	200	0	555	2,037
Prawns	Large portion	Cooked	150	0	31	520
Prawns, breaded	Large portion	Deep fried	150	0	5,751	682
Caviar, black	1 tablespoon	Raw	16	0	13	1,086
Tuna, fresh, bluefin	Medium portion	Raw	100	0	53	1,300
Tuna, canned in water	1 cup	Raw	154	0	14	433
Tuna, canned in oil	1 cup	Raw	146	0	3,917	295
Sardines, canned in tomato sauce	1 cup	Raw	89	1	109	1,507
Sardines, canned in oil	1 cup	Raw	149	0	5,280	2,205

Source: Data courtesy of USDA National Nutrient Database for Standard Reference Nutrition Data; https://nutritiondata.self.com.

FOOD	SERVING SIZE	Cooked	WEIGHT (g)	GLYCAEMIC load	OMEGA-6 (mg)	OMEGA-3 (mg)
Supermarket						
Pasta	Medium serving	Boiled	150	21	560	52
Rice, white	Medium serving	Boiled	150	24	98	20
Rice, wild	Medium serving	Boiled	164	16	195	156
Noodles	1 cup	Boiled	160	21	835	44
Bread, white	2 slices		50	16	304	34
Bread, wholewheat	2 slices		56	10	161	7
Sugar, white	1 teaspoon		4	3	0	0
Sugar, white			100	70	0	0
Flour			100	53	828	17
Crackers, wheat	4 crackers		50	17	1,350	70
Crackers, rye	4 crackers		44	16	156	20
Crisps	1 bag		28	11	3,010	53
Crisps	1 grab bag		60	23	6,100	120
Coca-Cola	1 can		330	11	0	0
Cake, sponge	1 slice		63	23	350	22
Pulses						
Kidney beans	½ can		200	12	212	164
Chickpeas	½ can		200	20	982	38
Baked beans	½ can		200	16	186	154

Source: Data courtesy of USDA National Nutrient Database for Standard Reference Nutrition Data; https://nutritiondata.self.com.

FOOD	SERVING SIZE	Cooked	WEIGHT (g)	GLYCAEMIC load	OMEGA-6 (mg)	OMEGA-3 (mg)
Cereal						
Fruit loops	1 cup		30	18	343	17
Frosties	1 cup		39	26	34	2
Corn flakes	1 cup		28	17	84	6
Alpen	½ cup		56	25	497	25
Nuts						
Almonds	1 small bag	Roasted	50	0	7,400	0
Cashews	1 small bag	Roasted	50	4	4,240	34
Macadamia	1 small bag	Roasted	50	0	645	102
Peanuts	1 small bag	Oil roasted	50	0	7,609	0
Peanut butter	2 tablespoons		32	0	4,709	26
Processed Meat						
Salami	4 slices	Chilled	100	0	1,940	420
Sausage, beef	1 sausage	Cooked	100	0	430	0
Sausage, pork	1 sausage	Cooked	70	0	2,430	80
Fast Food						
Double burger with cheese	1 item		400	27	10,353	1,564
Cheeseburger	1 item		133	16	1,818	164
Chicken Whopper	1 item		272	24	11,523	1,423
French fries	1 medium serving		177	26	1,310	31

Source: Data courtesy of USDA National Nutrient Database for Standard Reference Nutrition Data; https://nutritiondata.self.com.

FOOD	SERVING SIZE	Cooked	WEIGHT (g)	GLYCAEMIC load	OMEGA-6 (mg)	OMEGA-3 (mg)
Chocolate shake	Child size		267	39	507	42
Potato, hash browns	1 cup		150	27	6,800	527
12-inch cheese pizza	1 slice		95	15	1,563	188
14-inch pepperoni pizza	1 slice		85	13	2,482	299
Chicken, fried	1 piece, breast		140	6	2,800	143
Coleslaw	1 serving		112	6	4,840	634
Potato wedges	1 serving		134	22	2,303	107

Source: Data courtesy of USDA National Nutrient Database for Standard Reference Nutrition Data; https://nutritiondata.self.com.

Notes on Appendix 2

Natural omega 3:6 ratio of diet is 1:1 to 1:4.

Most Western diets have omega 3:6 ratio of 1:15 or more.

Aim to reduce your omega 3:6 ratio towards natural levels.

The glycaemic load of foods is dependent on portion size; i.e. 1 large baked potato = glycaemic load 29, 2 large baked potatoes = glycaemic load 58.

Step 5 of the plan encourages starting with a daily glycaemic load maximum of 100, then slowly reducing this to a daily load of 80, or lower if this is comfortable. Remember, when replacing carb-heavy foods do not avoid natural foods that contain saturated fats, such as meat and dairy products. Also, remember most vegetables have a low glycaemic load and a good omega profile.

COVID-19 and Obesity

*Pandemics: Diseases That Spread Over Whole
Countries or the Whole World*

In May 2020 the UK became Europe's biggest casualty of the corona-virus crisis. We were at the epicentre of the worst viral pandemic for a hundred years and had suffered more deaths than our neighbours. Accusations and politicization of the tragic mortality rate emerged. Did we start testing too late? Was the lockdown timing delayed for too long? Was there a shortage of PPE? But I wondered whether the main headlines and the news threads might be missing a point. Were we just more susceptible to this virus than other parts of the world? Was the near-death experience of our prime minister a totem of the vulnerability of our people?

Let's go back to the origin of coronavirus – China. If we are to believe their numbers, the early, extreme and prolonged lockdown of Wuhan province was a complete success. But curiously, despite many asymptomatic carriers taking the virus around the world, it did not take root within the rest of non-locked-down China.

The viral swarm descended on Italy, overrunning its health ser-vice and leading to a heartbreaking spike in deaths, particularly of the elderly. Spain followed and, just like in Italy, temporary morgues were constructed to house the corpses. Next came the UK. Despite a lockdown that was proportionally earlier than Italy's and a successful reconfiguration of the NHS to accommodate all COVID-19 patients, our death toll surpassed other, apparently less well-prepared, coun-tries. Why was this?

A clue comes from the contrasting experiences of the inhabitants of the wealthiest city in the world, New York, and the slum-dwellers of Mumbai in India.

The first case of coronavirus was reported in New York on 1 March. Despite lockdown from 22 March, and with the US having the most advanced healthcare system in the world, the number of deaths in the city is currently* 1,700 per million population.

In India, the first recorded case was on 30 January and Prime Minister Modi ordered a lockdown on 24 March. However, there was concern at the defencelessness of poor Indian slum-dwellers, whose crowded and unsanitary conditions precluded an effective lockdown. The famous Dharavi slum in Mumbai, featured in the film *Slumdog Millionaire*, has an estimated 1 million dwellers packed into an area of 2 square kilometres – like the population of Manchester living in an area the size of Hyde Park. How can you socially distance in these circumstances, when you are sharing a room with extended family and have to walk through crowded, narrow passages to access food, water and sanitation? In addition, these people were more likely to be malnourished and have weakened immune systems. The thought of the highly infectious coronavirus ripping through the vulnerable inhabitants of these shanty towns was shuddering. If it could do this to London and New York, what of these poor people?

However, the tragic news story of bodies piling up in slums did not materialize. There is no evidence that the predicted tsunami of death in India occurred. In fact, in Mumbai, the latest figures* show around 70 deaths per million.

So, why were affluent New Yorkers, with access to world-class healthcare, over twenty times more likely to succumb to the coronavirus infection, compared to residents of Mumbai? Why does COVID-19 seem to affect people living in affluent countries more severely than those living in poorer countries? Our understanding of how the virus works is beginning to unveil some nasty truths about the health of our population, and the populations of many developed Western countries.

We now know that the virus enters our bodies through a cellular door named ACE-2. This door is present in the mucous membranes of our nose and throat; it is also present in our lungs. It was initially thought that the effect of coronavirus in shutting down these protective doors in our lungs led to a severe and rapidly fatal pneumonia. This was the reason the Nightingale Hospitals were built in the UK – to support the thousands of people predicted to need ventilating, to give them a chance of surviving. However, there is emerging evidence that the ACE-2 door that COVID-19 has access to is also present in many other tissues of the body, including our heart

---

* At time of going to press (October 2020).

and blood vessels. And there is an increasing understanding of the damage that blocking the ACE-2 door in these locations can wreak.

Angiotensin-converting enzyme-2 (ACE-2) is a receptor protein that exists on the surface of human cells. Only recently discovered, in 2000, its actions are slowly being understood. It is known to help convert the enzyme angiotensin II to angiotensin 1,7.

Angiotensin II, the enzyme that ACE-2 disposes of, is produced as the final by-product of the renin-angiotensin system, a cascade of cellular reactions that occurs in response to either a fall in blood pressure (from bleeding or dehydration) or in response to a stressful situation (as a protection in case of a physical assault). Angiotensin II causes our blood vessels to constrict, our kidneys to reabsorb water and for us to become thirsty – all responses designed to raise our blood pressure and keep the oxygen pumping throughout our bodies. This is the traditional way that angiotensin II has been understood to work. However, recent evidence suggests that angiotensin II also has a more disruptive action. It is now known to increase the release of damaging inflammation throughout our bodies – the so-called 'cytokine storm' (thought to be the cause of the dreaded COVID pneumonia). The inflammatory response also exacerbates a condition called *oxidative stress.*

Oxidative Stress Corrodes the Body

Oxidative stress is crucial to understanding why some people suffer horribly with coronavirus while others hardly know they have had it. We may recognize oxidation in the world around us: it causes corrosion of metals, browning of fruits and rancidity of oily food (basically, food going off). It is a normal by-product of our internal metabolism. When our cells convert food energy into chemical energy they produce an excess of electrons. These are taken up by oxygen and converted into *superoxides* (oxygen with an electron attached) before finally being converted to harmless H_2O – water. Superoxides (and peroxides; think bleach), when in excess, can be damaging to the body by a process of oxidation. They are normally used by our inflammatory cells to kill invading bacteria and viruses (including COVID-19) but, unfortunately, if our bodies produce too much of them they cause cell damage and inflammation, leading to high rates of cancer, inflammatory conditions and heart disease. Sadly, there is evidence that a Western diet high in carbohydrates and infused with vegetable oils can turn up the production of super-oxides and lead to our own internal oxidation. Basically, a Western diet causes diseases through oxidative stress.

Our diet is encouraging our bodies to internally oxidize. We need an escape mechanism to minimize the damage. This is where ACE-2 comes in; it's our safeguard against the internal corrosion that our Western diet is producing.

ACE-2: The Primer Against Corrosion

When ACE-2 is working normally, angiotensin II is turned into angiotensin 1,7. This is part of our safeguarding mechanism against runaway inflammation and oxidative stress. Imagine it working like an anti-rust primer paint that you might apply to metal. Angiotensin 1,7 protects us by dilating our blood vessels (and reducing blood pressure) and its anti-inflammatory effects, including suppression of oxidative stress.

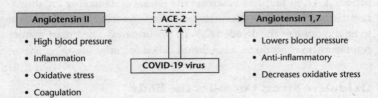

When coronavirus takes hold, the ACE-2 doors are locked, meaning our safeguarding mechanism malfunctions. Inflammation and oxidative stress take over. Essentially, the body's protective 'primer paint' has been removed. Emerging evidence, which fits into what intensivists are seeing on the ground, suggests that it is not just breathing that can be affected but, commonly, our blood vessels. When the ACE-2 located in our blood vessels are blocked by the virus, inflammation and oxidative stress act locally to cause blood clotting. This can lead to a cascade of blood clots forming throughout the body, causing heart attacks, embolisms, kidney failure and strokes. New York medics have noticed that many young patients who did not know they were infected with COVID-19 have been admitted with disabling or lethal strokes.

More ACE-2 in Obesity Means a Bigger Target for COVID-19

When we consume a diet rich in sugar and refined carbohydrates, our bodies have to work overtime to break down and store the excess calories that we consume. This leads to an overflow of oxidative stress. There is evidence that, to compensate for this, the amount of ACE-2 receptors is up-regulated in the presence of obesity, which means that more of these receptors are created to reduce any potential oxidative damage as a result of this typically Western diet. The same occurs with people suffering with heart failure and Type 2 diabetes – more ACE-2 receptors are made and we become more reliant on their inflammatory-stabilizing qualities.

Once COVID-19 blocks the ACE-2 receptors our real vulnerability is revealed. ACE-2 works like a dam, holding back a surging body of destructive water. In our case, it is holding back a surge of inflammatory and oxidative havoc. Once the virus breaks that protective shield, we feel the full force of its destructive power.

The role of ACE-2 in moderating our inflammatory response finally explains why people who suffer with obesity, and its related conditions such as cardiovascular disease and Type 2 diabetes, are at particular risk of succumbing to the virus. They are more reliant on the protective effect of ACE-2. Black and Asian people in Britain have higher rates variously of diabetes, hypertension and heart disease, which presumably means they highly rely on their ACE-2 dam. Our elderly population, even if apparently fit, already have high rates of oxidative stress (this is a natural side effect of aging), leaving them fewer reserves of the essential receptor when the virus hits.

Why are Slum-dwellers Less Affected by COVID?

Contrast this with the slum-dwellers of Dharavi. They rarely suffer with obesity or its associated diseases, such as Type 2 diabetes. When compared to affluent Westerners, they are much less reliant on ACE-2 to protect their bodies against oxidative stress because they do not generally consume Western foods (high in sugar, refined carbohydrates and fats). No excess ACE-2 receptors are required and therefore, when the virus does block them, less inflammatory damage to their body occurs. Coronavirus is rife in Dharavi at present – a recent report shows that 57 per cent of the residents of Dharavi now have antibodies to COVID (double the exposure compared to New York or London) – but the population does not have so many ACE-2 targets for the virus to hit and therefore far fewer will suffer lethal consequences of runaway oxidative stress and inflammation if they become infected.

The Two Pandemics

The coronavirus pandemic shocked the world. It engulfed us suddenly and caused many tragic deaths (over 1 million to date*). But there was another pandemic already present, one that has crept up on us slowly over thirty years: the obesity pandemic. A disease that has caused many more early deaths than coronaviruses (4 million excess deaths per year, every year for the last twenty years).

The UK is the sick man of Europe when it comes to obesity. The US is even sicker. The Western diet leaves our bodies exposed to the full power of COVID-19. Maybe by understanding this we can change the way we eat to protect ourselves for the future. If we consumed less of these problematic foods, snacked less and cooked with healthy, fresh food instead (as described in Part Three), we could achieve this.

COVID-19 taught us that, as a people, we can come together, once we know the health risks, to dramatically change our behaviour and the way we live. COVID-19 is going to be with us for many years to come, maybe mutating and hitting us with repeated surges of infection. By aggressively tackling the obesity pandemic once and for all the government will not only improve the health, quality of life and longevity of its people, it will also protect our population against future coronavirus outbreaks. In the meantime we can, as individuals, make these dietary changes and decrease our risks of the deadly consequences of this virus.

*

In April 2020, I attended the funeral of my good friend (and former patient) Panny. He had tragically died from coronavirus. He leaves behind a loving family and young son. Rest in peace, my friend.

---

* At time of going to press (October 2020).

REFERENCES

1 Metabology for Beginners

1 USDH (1998). Clinical guidelines on the identification, evaluation, and treatment of overweight and obesity in adults: the evidence report. National Institute of Health (NIH) Publication, No. 98-4083, September.

2 R. Bailey (2018). *Evaluating Calorie Intake for Population Statistical Estimates (ECLIPSE) Project*, February. Office for National Statistics, Data Science Campus.

3 P. Miller (2015). The United States food supply is not consistent with dietary guidance: evidence from an evaluation using the Healthy Eating Index-2010. *J Acad Nutr Diet*, 115(1), January, 95–100.

4 J. Speakerman (2004). The functional significance of individual variation in basal metabolic rate. *Physiol Biochem Zool*, 77(6), 900–915.

5 G. Koepp (2016). Chair-based fidgeting and energy expenditure. *BMJ Open Sport Exerc Med*, 2(1).

6 E. Sims and E. Horton (1968). Endocrine and metabolic adaptation to obesity and starvation. *Am J Clin Nutr*, 21(12), December, 1455–70.

7 R. Leibel et al. (2000). Effects of changes in body weight on carbohydrate metabolism, catecholamine excretion, and thyroid function. *Am J Clin Nutr*, 71(6), June, 1421–32.

8 A. Harris et al. (2006). Weekly changes in basal metabolic rate with eight weeks of overfeeding. *Obesity (Silver Spring)*, 14(4), April, 690–95.

9 C. Weyer et al. (2001). Changes in energy metabolism in response to 48 h of overfeeding and fasting in Caucasians and Pima Indians. *Int J Obes Relat Metab Disord*, 25(5), May, 593–600.

10 A. Keys et al. (1950). *The Biology of Human Starvation*, Vol. 1. Minneapolis, University of Minnesota Press.

11 R. Leibel et al. (1995). Changes in energy expenditure resulting from altered body weight. *N Eng J Med*, 332(10), March, 621–8; S. Roberts and I. Rosenberg (2006). Nutrition and aging: changes in the regulation of energy metabolism with aging. *Physiol Rev*, 86(2), April, 651–67.

12 A. Evans et al. (2016). Drivers of hibernation in the brown bear. *Frontiers in Zoology*, 13, February, article no. 7.

13 R. Keesey (1997). Body weight set-points: determination and adjustment. *J Nutr*, 127(9), September, 1875S–1883S.

2 The Sacred Cow

1 B. Levin et al. (1989). Initiation and perpetuation of obesity and obesity resistance in rats. *Am J Physiol Regul Integr*, 256 (3, Pt 2), R766–71.

2 M. Butovskaya et al. (2017). Waist-to-hip ratio, body-mass index, age and number of children in seven traditional societies. *Sci Rep*, 7(1), May, 1622.

3 M. Ashwell et al. (2014). Waist-to-height ratio is more predictive of years of life lost than body mass index. *PLoS One*, 9(9), September.

4 V. Eshed et al. (2010). Paleopathology and the origin of agriculture in the Levant. *Am J Phys Anthropol*, 143(1), September, 121–33.

5 World Health Organization (2016). *Global Health Observatory Data*.

6 J. Wardle and D. Boniface (2008). Changes in the distributions of body mass index and waist circumference in English adults, 1993/1994 to 2002/2003. *Int J Obes (Lond)*, 32(3), March, 527–32.

7 Reuters/Ipsos (2012). Ipsos online poll of 1,143 adults, 7–10 May. Reuters.

8 C. Haworth et al. (2008). Childhood obesity: genetic and environmental overlap with normal-range BMI. *Obesity*, 16(7), July, 1585–90.

9 Q. Xia and S. F. Grant (2013). The genetics of human obesity. *Ann N Y Acad Sci*, 1281, April, 178–90.

10 B. Gascoigne (2001). Retrieved 2018, from HistoryWorld: www.history world.net.

11 J. Terrell (ed.) (1988). *Von den Steinen's Marquesan Myths*, translated by Marta Langridge. Canberra: Target Oceania/*Journal of Pacific History*.

12 R. O'Rourke (2015). Metabolic thrift and the genetic basis of human obesity. *Ann Surg*, 259(4), April, 642–8.

13 J. Neel (1962). Diabetes mellitus: a 'thrifty' genotype rendered detrimental by 'progress'? *Am J Hum Genet*, 14, December, 353–62.

14 World Health Organization (2016). *Global Health Observatory Data*.

15 P. Manning (1992). 'The Slave Trade: The Formal Demography of a Global System', in J. E. Inikori and S. L. Engerman (eds), *The Atlantic Slave Trade*. Durham, NC: Duke University Press.

16 A. Quasim et al. (2018). On the origin of obesity: identifying the biological, environmental and cultural drivers of genetic risk among human populations. *Obes Rev*, 19(2), February, 121–49.

17 Y. Wang and M. Beydoun (2007). The obesity epidemic in the United States – gender, age, socioeconomic, racial/ethnic, and geographic characteristics: a systematic review and meta-regression analysis. *Epidemiol Rev*, 29, 6–28; Centers for Disease Control and Prevention (CDC) (2012). *National Health and Nutrition Examination Survey, NHANES 2011–2012 Overview*. National Center for Health Statistics.

18 S. van Dijk et al. (2015). Epigenetics and human obesity. *Int J Obes*, 39(1), 85–97.

19 Z. Stein and M. Susser (1975). The Dutch famine, 1944–1945, and the reproductive process. I. Effects on six indices at birth, *Pediatric Research*, 9, February 70–76.

20 M. Hult et al. (2010). Hypertension, diabetes and overweight: looming legacies of the Biafran famine. *PLoS One*, 5(10), October, e13582.

21 B. Weinhold (2006). Epigenetics: the science of change. *Environ Health Perspect*, 114(3), March, A160–A167.

22 I. Ehrenreich and D. Pfennig (2016). Genetic assimilation: a review of its potential proximate causes and evolutionary consequences. *Ann Bot*, 117(5), April, 769–79.

23 A. Samuelsson et al. (2008). Diet-induced obesity in female mice leads to offspring hyperphagia, adiposity, hypertension, and insulin resistance: a novel murine model of developmental programming. *Hypertension*, 51(2), February, 383–92.

24 A. Kubo et al. (2014). Maternal hyperglycemia during pregnancy predicts adiposity of the offspring. *Diabetes Care*, 37(11), November, 2996–3002.

25 A. Sharma et al. (2005). The association between pregnancy weight gain and childhood overweight is modified by mother's pre-pregnancy BMI. *Pediatr Res*, 58, 1038.

26 F. Guenard et al. (2013). Differential methylation in glucoregulatory genes of offspring born before vs. after maternal gastrointestinal bypass surgery. *Proc Natl Acad Sci USA*, 110(28), July, 11439–44.

27 R. Waterland and R. Jirtle (2003). Transposable elements: targets for early nutritional effects on epigenetic gene regulation. *Mol Cell Biol*, 23(15), August, 5293–300.

28 Waterland and Jirtle (2003). Transposable elements.

3 Dieting and the Biggest Losers

1 E. Fothergill et al. (2016). Persistent metabolic adaptation for 6 years after 'The Biggest Loser' competition. *Obesity (Silver Spring)*, 24(8), August, 1612–19.

2 H. Yoo et al. (2010). Difference of body compositional changes according to the presence of weight cycling in a community-based weight control program. *J Korean Med Sci*, 25(1), January, 49–53.

3 S. Dankel et al. (2014). Weight cycling promotes fat gain and altered clock gene expression in adipose tissue in C57BL/6J mice. *Am J Physiol Endocrinol Metab*, 306(2), January, E210–24.

4 J. Speakerman et al. (2004). The functional significance of individual variation in basal metabolic rate. *Physiol Biochem Zool*, 77(6), November–December, 900–915.

5 L. Arone et al. (1995). Autonomic nervous system activity in weight gain and weight loss. *Am J Physiol*, 269(1, Pt 2), R222–5.

6 K. O'Dea et al. (1982). Noradrenaline turnover during under- and overeating in normal weight subjects. *Metabolism*, 31(9), September, 896–9; S. Welle et al. (1991). Reduced metabolic rate during beta-adrenergic blockade in humans. *Metabolism*, 40(6), June, 619–22; A. Thorp and M. Schlaich (2015). Relevance of sympathetic nervous system activation in obesity and metabolic syndrome. *J Diabetes Res*, 2015, 341583.

7 J. Grundlingh et al. (2011). 2,4-dinitrophenol (DNP): a weight loss agent with significant acute toxicity and risk of death. *J Med Toxicol*, 7(3), September, 205–12.

4 Why We Eat

1 D. Cummings et al. (2002). Plasma ghrelin levels after diet-induced weight loss or gastric bypass surgery. *N Eng J Med*, 346(21), May, 1623–30.

2 P. Sumithran et al. (2011). Long-term persistence of hormonal adaptations to weight loss. *N Eng J Med*, 365(17), October, 1597–1604.

3 J. Cirello and J. Moreau (2013). Systemic administration of leptin potentiates the response of neurons in the nucleus of the solitary tract to chemoreceptor activation in the rat. *Neuroscience*, 229, January, 89–99.

4 Y. Zhang et al. (1994). Positional cloning of the mouse obese gene and its human homologue. *Nature*, 372(6505), December, 425–32.

5 C. Montague et al. (1997). Congenital leptin deficiency is associated with severe early-onset obesity in humans. *Nature*, 387(6636), June, 903–8.

6 S. Heymsfield et al. (1999). Recombinant leptin for weight loss in obese and lean adults: a randomized, controlled, dose-escalation trial. *JAMA*, 282(16), October, 1568–75.

5 The Glutton

1 F. Chehab (2014). 20 years of leptin: leptin and reproduction: past mile-stones, present undertakings, and future endeavours. *J Endocrinol*, 223(1), October, T37–48.

2 Chehab (2014). 20 years of leptin.

3 R. Lustig (2013). *Fat Chance: Beating the odds against sugar, processed food, obesity and disease*. New York: Hudson Street Press.

4 S. Ramirez and M. Claret (2015). Hypothalamic ER stress: a bridge between leptin resistance and obesity. *FEBS Lett*, 589(14), June, 1678–87.

5 R. Lustig et al. (2004). Obesity, leptin resistance and the effects of insulin reduction. *Int J Obes Relat Metab Discord*, 28(10), October, 1344–8.

6 B. Wisse and M. Schwartz (2009). Does hypothalamic inflammation cause obesity? *Cell Metab*, 10(4), October, 241–2.

7 I. Nieto-Vazquez et al. (2008). Insulin resistance associated to obesity: the link TNF-alpha. *Arch Physiol Biochem*, 114(3), July, 183–94.

8 Chehab (2014). 20 years of leptin.

9 J. Wang et al. (2001). Overfeeding rapidly induces leptin and insulin resistance. *Diabetes*, 50(12), December, 2786–91.

7 The Master Chef

1 R. Dawkins (1989). *The Selfish Gene*, 2nd edn. Oxford: Oxford University Press.

2 L. C. Aiello and P. Wheeler (1995). The expensive-tissue hypothesis: the brain and the digestive system in human and primate evolution. *Current Anthropology*, 36(2), April, 199–221.

3 F. Berna et al. (2012). Microstratigraphic evidence of in situ fire in the Acheulean strata of Wonderwerk Cave, Northern Cape province, South Africa. *PNAS*, 109(20), May, E1215–20.

4 C. Koebnick et al. (1999). Consequences of a long-term raw food diet on body weight and menstruation: results of a questionnaire survey. *Ann Nutr Metab*, 43(2), 69–79.

5 I. Olalde et al. (2014). Derived immune and ancestral pigmentation alleles in a 7,000-year-old Mesolithic European. *Nature*, 507(7491), March, 225–8.

6 D. Bramble and D. Lieberman (2004). Endurance running and the evolution of *Homo*. *Nature*, 432 (7015), November, 345–52.

7 P. Williams (2007). Nutritional composition of red meat. *Nutrition and Dietetics*, 64(4), August, 113–19.

8 P. Clayton (2009). How the mid-Victorians worked, ate and died. *Int J Environ Res Public Health*, 6(3), March, 1235–53.

8 The Heart of the Matter

1 US Department of Agriculture Economic Research Service – Food Availability; Statistical Abstract of the United States. US Government Printing Office, 763.

2 J. Yudkin (1972). *Pure, White and Deadly: How sugar is killing us and what we can do to stop it.* London: Davis-Poynter; reissue London: Penguin Books, 2012.

3 R. McGandy et al. (1967). Dietary fats, carbohydrates and atherosclerotic vascular disease. *N Eng J Med*, 277(4), July, 186–92.

4 C. Kearns (2016). Sugar industry and coronary heart disease research: a historical analysis of internal industry documents. *JAMA Intern Med*, 176(11), November, 1680–85.

5 A. Keys (1980). *Seven Countries: A multivariate analysis of death and coronary heart disease.* Cambridge, MA: Harvard University Press.

6 Keys (1980). *Seven Countries.*

7 N. Teicholz (2014). *The Big Fat Surprise: Why butter, meat and cheese belong in a healthy diet.* New York: Simon & Schuster.

8 R. H. Lustig (2013). *Fat Chance: The hidden truth about sugar, obesity and disease.* London: Fourth Estate.

9 Teicholz (2014). *The Big Fat Surprise*, p. 101.

10 E. Steele et al. (2016). Ultra-processed foods and added sugars in the US diet: evidence from a nationally representative cross-sectional study. *BMJ Open*, 6(3), March.

11 P. Clayton (2009). How the mid-Victorians worked, ate and died. *Int J Environ Res Public Health*, 6(3), March, 1235–53; J. E. Bennett et al. (2015). The future of life expectancy and life expectancy inequalities in England and Wales: Bayesian spatiotemporal forecasting. *The Lancet*, 386(9989), July, 163–70.

9 The Omega Code

1 D. Arnold (2010). British India and the 'Beriberi Problem', 1798–1942. *Med Hist*, 54(3), July, 295–314.

2 A. Hawk (2006). The great disease enemy, Kak'ke (beriberi) and the Imperial Japanese Army. *Military Medicine*, 171(4), 333–9.

3 N. Raizman (2004). Review of S. R. Bown, *Scurvy: How a Surgeon, a Mariner, and a Gentleman Solved the Greatest Medical Mystery of the Age of Sail* (New York: St Martin's Press, 2003). *J Clin Invest*, 114(12), December, 1690.

4 J. Lind (1753). *A Treatise of the Scurvy.* Edinburgh: A. Kincaid and A. Donaldson.

5 S. Allport (2006). *The Queen of Fats*. Berkeley, CA: University of California Press.

6 C. E. Ramsden et al. (2013). Use of dietary linoleic acid for secondary prevention of coronary heart disease and death: evaluation of recovered data from the Sydney Diet Heart Study and updated meta-analysis. *BMJ*, 346, February, e8707.

7 A. P. Simopoulos (2004). Omega-6/omega-3 essential fatty acid ratio and chronic diseases. *Food Reviews International*, 20(1), 77–90.

8 H. Freitas et al. (2017). Polyunsaturated fatty acids and endocannabinoids in health and disease. *Nutr Neurosci*, 21(1), July, 1–20.

9 A. P. Simopoulos (2016). An increase in the omega-6/omega-3 fatty acid ratio increases the risk for obesity. *Nutrients*, 8(3), March, 128.

10 S. Banni and V. Di Marzo (2010). Effect of dietary fat on endocannabinoids and related mediators: consequences on energy homeostasis, inflammation and mood. *Mol Nutr Food Res*, 54(1), January, 82–92; I. Matias and V. Di Marzo (2007). Endocannabinoids and the control of energy balance. *Trends Endocrinol. Metab*, 18(1), January–February, 27–37.

11 Allport (2006). *The Queen of Fats*.

12 A. Evans (2016). Drivers of hibernation in the brown bear. *Frontiers in Zoology*, 13, February, article no. 7.

13 T. Ruf and W. Arnold (2008). Effects of polyunsaturated fatty acids on hibernation and torpor: a review and hypothesis. *Am J Physiol Regul Integr Comp Physiol*, 294(3), March, R1044–52; D. Munro and D. W. Thomas (2004). The role of polyunsaturated fatty acids in the expression of torpor by mammals: a review. *Zoology*, 107(1), 29–48.

14 G. L. Florant (1998). Lipid metabolism in hibernators: the importance of essential fatty acids. *Amer Zool*, 38, 331–40.

15 V. Hill and G. L. Florant (2000). The effect of a linseed oil diet on hibernation in yellow-bellied marmots (*Marmota flaviventris*). *Physiol Behav*, 68(4), February, 431–7.

16 Allport (2006). *The Queen of Fats*.

10 The Sugar Roller Coaster

1 P. Evans and R. Lynch (2003). Insulin as a drug of abuse in body building. *Br J Sports Med*, 37(4), August, 356–7.

2 R. Henry et al. (1993). Intensive conventional insulin therapy for type II diabetes. Metabolic effects during a 6-mo outpatient trial. *Diabetes Care*, 16(1), January, 21–31.

3 R. H. Lustig et al. (2003). Suppression of insulin secretion is associated

with weight loss and altered macronutrient intake and preference in a sub-set of obese adults. *Int J Obes Relat Metab Disord*, 27(2), February, 219–26.

4 C. S. Lieber et al. (1991). Perspectives: do alcohol calories count? *Am J Clin Nutr*, 54(6), 976–82.

5 P. Suter (2005). Is alcohol consumption a risk factor for weight gain and obesity? *Crit Rev Clin Lab Sci*, 42(3), 197–227.

6 L. Cordain et al. (1997). Influence of moderate daily wine consumption on body weight regulation and metabolism in healthy free-living males. *J Am Coll Nutr*, 16(2), April, 134–9.

7 A. Arif and J. Rohrer (2005). Patterns of alcohol drinking and its associ-ation with obesity: data from the Third National Health and Nutrition Survey 1988–1994. *BMC Public Health*, 5, December, 126.

8 T. Stalder et al. (2010). Use of hair cortisol analysis to detect hypercorti-solism during active drinking phase in alcohol-dependent individuals. *Biol Psychol*, 85(3), December, 357–60.

11 The French Paradox

1 P. MacLean and R. Batterham et al. (2017). Biological control of appetite: a daunting complexity. *Obesity (Silver Spring)*, 25(1), March, S8–S16.

2 D. Treit and M. L. Spetch (1986). Caloric regulation in the rat: control by two factors. *Physiology & Behavior*, 36(2), 311–17.

13 The Fat of the Land

1 M. Sladek et al. (2016). Perceived stress, coping, and cortisol reactivity in daily life: a study of adolescents during the first year of college. *Biol Psychol*, 117, May, 8–15; A. Bhende et al. (2010). Evaluation of physiologic-al stress in college students during examination. *Biosc Biotech Res Comm*, 3(2), December, 213–16.

2 S. Gropper et al. (2012). Changes in body weight, composition, and shape: a 4-year study of college students. *Appl Physiol Nutr Metab*, 37(6), 1118–23.

3 L. Dinour et al. (2012). The association between marital transitions, body mass index, and weight: a review of the literature. *J Obes*, 2012(294974), May.

4 T. Robles and J. Kiecolt-Glaser (2003). The physiology of marriage: path-ways to health. *Physiol Behav*, 79(3), August, 409–16.

5 P. B. Gray et al (2004). Social variables predict between-subject but not day-to-day variation in the testosterone of US men. *Psychoneuroendocrin-ology*, 29(9), October, 1153–62; E. Barrett et al. (2015). Women who are married or living as married have higher salivary estradiol and progester-one than unmarried women. *Am J Hum Biol*, 27(4), July–August, 501–7.

6 B. Leeners et al. (2017). Ovarian hormones and obesity. *Hum Reprod Update*, 23(3), May, 300–321.

7 J. Cipolla-Neto et al. (2014). Melatonin, energy metabolism, and obesity: a review. *J Pineal Res*, 56(4), May, 371–81.

8 Cipolla-Neto et al. Melatonin, energy metabolism, and obesity.

9 M. Mankowska et al. (2017). Confirmation that a deletion in the POMC gene is associated with body weight of Labrador Retriever dogs. *Res Vet Sci*, 112, June, 116–18.

10 H. Eicher-Miller et al. (2012). Contributions of processed foods to dietary intake in the US from 2003–2008: a report of the Food and Nutrition Science Solutions Joint Task Force of the Academy of Nutrition and Dietetics, American Society for Nutrition, Institute of Food Technologists, and International Food Information Council. *J Nutr*, 142(11), November, 2065S–2072S.

11 C. Monteiro et al. (2018). Household availability of ultra-processed foods and obesity in nineteen European countries. *Public Health Nutr*, 21(1), January, 18–26.

14 Prepare to Do It Yourself

1 Z. T. Segal et al. (2012). *Mindfulness-Based Cognitive Therapy for Depression*, 2nd edn. New York: The Guilford Press.

15 Eat More, Rest More

1 M. Walker (2017). *Why We Sleep: Unlocking the power of sleep and dreams*. London: Penguin Books.

16 Your Personal Blue Zone

1 R. De Souza et al. (2015). Intake of saturated and trans unsaturated fatty acids and risk of all cause mortality, cardiovascular disease, and type 2 diabetes: systematic review and meta-analysis of observational studies. *BMJ*, 351, August, h3978.

2 H. Pontzer et al. (2012). Hunter-gatherer energetics and human obesity. *PLoS One*, 7(7), July, e40503.

Appendix 1: The Cholesterol Debate

1 M. Gladwell (2000). *The Tipping Point: How little things can make a big difference*. London: Little, Brown.

2 A. Keys (1980). *Seven Countries: A multivariate analysis of death and coronary heart disease*. Cambridge, MA: Harvard University Press.

3 C. Kearns et al. (2016). Sugar industry and coronary heart disease research: a historical analysis of internal industry documents. *JAMA Intern Med*, 176(11), November, 1680–85.

4 S. Hamley (2017). The effect of replacing saturated fat with mostly n-6 polyunsaturated fat on coronary heart disease: a meta-analysis of randomised controlled trials. *Nutr J*, 16(1), May, article no. 30; S. Berger et al. (2015). Dietary cholesterol and cardiovascular disease: a systematic review and meta-analysis. *Am J Clin Nutr*, 102(2), August, 276–94.

5 R. De Souza et al. (2015). Intake of saturated and trans unsaturated fatty acids and risk of all cause mortality, cardiovascular disease, and type 2 diabetes: systematic review and meta-analysis of observational studies. *BMJ*, 351, August, h3978.

6 P. Siri and R. Krauss (2005). Influence of dietary carbohydrate and fat on LDL and HDL particle distributions. *Curr Atheroscler Rep*, 7(6), November, 455–9; P. Siri-Tarino et al. (2010). Saturated fat, carbohydrate, and cardiovascular disease. *Am J Clin Nutr*, 91(3), March, 502–9.

7 J. Durstine et al. (2002). Lipids, lipoproteins, and exercise. *J Cardiopulm Rehabil*, 22(6), November–December, 385–98.

8 F. Sacks et al. (2017). Dietary fats and cardiovascular disease: a presidential advisory from the American Heart Association. *Circulation*, 136(3), July, e1–e23.

9 R. Krauss (1995). Dense low density lipoproteins and coronary artery disease. *Am J Cardiol*, 75(6), February, 53B–57B.

10 World Health Organization (2003). *Diet, Nutrition and the Prevention of Chronic Diseases*. WHO Technical Report Series, 916, 10, 88.

11 De Souza et al. (2015). Intake of saturated and trans unsaturated fatty acids.

Appendix 2: Glycaemic Load and Omega-3 to Omega-6 Ratio of Common Foods

1 S. A. Khan (2017). Comparative study of fatty-acid composition of table eggs from Jeddah food market and effect of value addition in omega-3 bio-fortified eggs. *Saudi J Biol Sci*, 24(2), 929–35.

2 Khan (2017). Comparative study of fatty-acid composition of table eggs.

GLOSSARY

adenosine triphosphate (ATP) – ATP is the chemical found within the cells of all living organisms on Earth. It acts as an energy currency that cells can understand and use. ATP stores the energy that is released when food is broken down and transports it to the area of the cell requiring energy for building and repair.

ATP batteries – ATP molecules act like mini cellular batteries, charging up (on food) and discharging their energy in other parts of the cell.

autonomic nervous system (ANS) – The autonomic nervous system describes a part of our nervous system that is not under our conscious control (it is automatic). It is split into two parts: the *sympathetic nervous system (SNS)* and the *parasympathetic nervous system (PNS)*. These two systems work either to optimize physical activity (in times of danger) or to preserve energy.

basal metabolic rate (BMR) – BMR describes the amount of energy that the body uses when at rest, including the energy required for cellular chemical reactions (building and repair), temperature control, breathing and heart rate.

dinitrophenol (DNP) – A chemical substance that causes the release of the energy stored in *ATP* to thermal (heat) rather than to chemical energy.

epigenetics – The study of the way inherited traits from DNA can be altered during pregnancy and early childhood in response to the environment.

ghrelin – A hormone produced by the stomach (and upper GI tract) that acts (via the *hypothalamus*) to produce a voracious appetite and food-seeking behaviour. Ghrelin increases in response to starvation (and dieting) and decreases after eating.

GLP-1 – Glucagon-like peptide-1 is a hormone released by the small bowel when food is being digested. It acts (via the *hypothalamus*) to increase satiety and is part of the signal to stop eating. It also acts to improve the efficiency of *insulin*.

hypothalamus – The pea-sized gland within the brain that is responsible for processing incoming sensory information such as the state of hydration and nutrition. In response to incoming signals it determines thirst, hunger and metabolic rate.

insulin – A hormone produced by the pancreas in response to food, particularly carbohydrates. It works to clear excess glucose (sugar) from the blood by opening the channels in the cells that suck in glucose.

leptin – A hormone produced by fat cells. Leptin acts as the 'master regulator' of body weight. When fat accumulates, leptin levels rise. This signals to the *hypothalamus* that enough energy is stored, resulting in an increase in metabolism and decreased appetite. When there is less fat, leptin levels fall, resulting in the hypothalamus increasing appetite and decreasing metabolism.

leptin resistance – The presence of very high levels of *leptin* that are not sensed by the *hypothalamus*. The leptin signal is blocked by *insulin* and *TNF-alpha* (inflammation). Despite there being a high level of fat in the body, the hypothalamus does not sense this and therefore does not correct it.

metabolic adaptation – Alterations in the amount of energy expended in response to the amount of energy consumed in order to defend a *weight set-point* and stop extreme weight fluctuations. The *metabolic rate* shifts downwards in response to calorie restriction (opposing extreme weight loss), and upwards in response to excess calories (opposing extreme weight gain).

metabolic rate – In this book metabolic rate refers to resting metabolic rate, i.e. the amount of energy required by the body to function while at rest.

micro-batteries – In this book the term micro-batteries is used to describe the function of *ATP*, small cellular chemicals that constantly charge and then discharge their energy, acting like mobile chargers.

negative feedback system – A system that is designed to maintain order by automatically correcting changes away from the pre-set desired equilibrium.

obesogenic – Something that promotes obesity.

omega fatty acids – The term refers to the two polyunsaturated fatty acids omega-3 and omega-6. Omega fatty acids are essential for cellular health. Humans are unable to make them and therefore foods that contain them should be part of a healthy diet.

parasympathetic nervous system (PNS) – Part of the *autonomic nervous system*. The PNS promotes energy conservation by decreasing pulse and blood pressure.

peptide-YY (PYY) – A hormone originating in the small intestine that is released after food is sensed within the bowel. It acts on the *hypothalamus* to promote feelings of satiety, or fullness, and forms part of the signal to stop eating.

sympathetic nervous system (SNS) – Part of the *autonomic nervous system*. The SNS triggers the fight or flight response to danger, enhancing strength, speed, and clarity of thought by increasing blood (and oxygen) flow to the muscles and brain.

thermogenesis – The process by which cellular energy, in the form of *ATP*, is converted to thermal energy (heat) rather than to chemical or mechanical energy.

TNF-alpha – Tumour necrosis factor-alpha is a protein that is released by inflammatory cells in response to the threat of attack (real or perceived). It helps to stimulate the inflammatory response seen in both infection and in autoimmune disease.

weight-control centre – A term used in this book to describe the *hypothalamus*.

weight set-point – The term refers to the weight that the body senses is safest for its survival and reproduction. The weight set-point is determined by genetic, *epigenetic* and environmental factors.

BIBLIOGRAPHY

Allport, Susan, *The Queen of Fats* (Berkeley, CA: University of California Press, 2006)

Briffa, John, *Escape the Diet Trap* (London: Fourth Estate, 2012)

Buettner, Dan, *The Blue Zones* (Washington DC: National Geographic, 2008)

Davis, William, *Wheat Belly* (London: HarperThorsons, 2014)

Guyenet, Stephan, *The Hungry Brain* (London: Vermilion, 2017)

Hoffmann, Peter, *Life's Ratchet* (New York: Basic Books, 2012)

Lewis, David, and Margaret Leitch, *Fat Planet* (London: Random House Books, 2015)

Lustig, Robert, *Fat Chance* (London: Fourth Estate, 2014)

Moalem, Sharon, *Survival of the Sickest* (London: HarperCollins, 2008)

Nesse, Randolph, and George Williams, *Why We Get Sick* (New York: Vintage Books, 1996)

Pollan, Michael, *In Defence of Food* (London: Allen Lane, 2008)

Sisson, Mark, *The Primal Blueprint* (London: Ebury Press, 2012)

Taubes, Gary, *The Case against Sugar* (London: Portobello Books, 2017)

Teicholz, Nina, *The Big Fat Surprise* (London: Scribe, 2014)

Wrangham, Richard, *Catching Fire* (London: Profile Books, 2009)

ACKNOWLEDGEMENTS

As I stated at the start of this book, the inspiration for writing it came from the many patients that I listened to and befriended in my clinics. They are the reason I wrote this book and it is to them that I am truly thankful for their early encouragement. I hope this book will repay the faith they have in me. A special mention should go to Jak, the first patient that I operated on, and in particular to his mother Dina, and to her family and community, for their support. Panny, Jerry, Satish, Alicia, Elisa, Yenti, Norma and all of you, thanks for your support.

Obviously, a book takes time to write and preparation to write it. Thanks to my wonderful loving girls at home, Rina, Jessica and Hannah, for your support and constant good humour. Thanks to Mum, who read the first draft and, unsurprisingly, loved it; Dad, who did the DIY in the study (at the age of eighty); and Richard and Sarah (and their families) for their support.

The book would not have been possible without my eighteen-month sabbatical from NHS duties, so I am indebted to Richard Cohen and Sarah Shaw at UCLH for facilitating this – I hope I have repaid that trust with this book.

I am also indebted to the surgeons that inspired and trained me, particularly David McLean, Don Menzies, Abrie Botha and Kesava Mannur. My research thesis would not have been possible without the intervention of Professor Norman Williams – thank you. Appreciation to my friends in the lab: Sri, Etsuro, Scotty, David Evans and, obviously, Charlie Knowles (now Professor).

Thanks to the support of the bariatric team, past and present, at UCLH, particularly my friends Marco Adamo, Rachel Batterham,

Mo, Majid, Naim, Andrei, Andrea Pucci, Himender Makker, Muntzer Mughal, Billy White, James Holding, Jackie Doyle, Kate Waller, Lise, Alison and Dr Maan Hasan, the safest and most interesting anaesthetist in London. Not to forget the admin team, past and present, at UCH: Jason Willis, Andreas Mann, Jade O'Connell, Maleika Pitterson . . . all of you – thanks.

Thanks to the research team I am currently working with at UCLH to investigate the effect of dieting on resting metabolic rate: Belinda Dury, Jessica Mok, Rob Stephens and Rachel Batterham (again).

A special mention should go to Natalie Cole – the highly efficient manager of my private practice in Harley Street . . . keep those twenty to thirty emails a day coming; I need the reminders!

I am grateful for my friends and colleagues in the United Arab Emirates for their unwavering support and their enthusiasm for this book: Rola Ghali, Mike Stroud, Alaa, Dr Bilal, Samer, Medhat, Fahmeeda, and Chandni Sharma.

Appreciation to Dr John Briffa, author and GP, for his advice to 'write like you talk' – thanks, John. And to Kevin Harvey for reading an early manuscript and giving me invaluable advice.

I am truly grateful to Elizabeth Sheinkman, my literary agent, for her fantastic enthusiasm for the book, and for recommending (from many offers) Penguin Life as publishers. At Penguin, I am indebted to Venetia Butterfield and Marianne Tatepo for moulding the book into the finished product it now is, and finally to Jane Robertson, my copy-editor, for adding clarity and focus to my writing.

INDEX

Page references in *italic* indicate Figures.

neo-Darwinism 54
neo-Lamarckism 54
Nestlé 150
New England Journal of Medicine 141,
    147–8
nicotinamide adenine dinucleotide
    (NADH) 205–6
night work, weight gain with
    226–7, 252–3
nutritional research
  heart disease correlations with fat
    140, 141–2, *143*, 145–53, 171
  weaknesses and biases 142–5
nutritionism 211n
nuts 114, 119, 123, *127*, *156*, 222,
    258, 261
  glycaemic load *291*

obesity
  blue zone lack of problems with
    256
  and BMI 37–8
  and calorie consumption 12–15,
    13, *14*, 17
  dietary deficiency as a trigger/
    driver for 170–95
  and 'diseases of civilization'
    155–9, *155*
  and energy balance xiii, 5, 6–9,
    8, 12–15
  and environmental change 32–3,
    34–42, 43–4, 47, 61, 193–4, 224,
    228–30
  epidemic 158–60, 171–5, *172*, *174*
  and epigenetics 51–61
  European rates 44
  food quality, weight set-point and
    29, 33, 38–9, 207, 230, 264–5,
    270, 275
  and free will 42–3

generational shift in risk 58–9
genetic predisposition to 33–4,
    35, 42–62
and Gulf States 44, 50–51, 54–5
hidden obesity genes 46
identical twin study 42–3
league table 44
leptin as hoped-for cure 87–8
and leptin resistance 92–101, *99*,
    186, 187, 223, 275
lifestyle-choice view of 275
link with intake of vegetable oils
    in USA *190*
listening to patients with ix,
    xii–xiv, 59, 106–7
and maternal blood sugar during
    pregnancy 55–6
and melatonin deficiency 227–8
Middle Eastern women's rates
    160, 267
and negative feedback 5, 9–12,
    15–30
obesogenic traits in offspring,
    and maternal over-nutrition in
    pregnancy 55–8
and omega-3 to omega-6 ratio
    183–95, 256
and Pacific Islanders 44, 45–6
and Pima tribe 47
as a pro-inflammatory
    condition 97
and processed food *see* processed/
    Western-type foods
programme to combat *see* weight
    set-point-based programme for
    losing weight
and radiation from atomic
    testing 49
and reproductive fitness
    hypothesis 46–7

weight-cycling dieting 65-7, *67*

weight gain

and calorie consumption 12-15,
13, *14*, 17

of cattle 32-6, 60-61

and insulin therapy 197-8

and leptin resistance 92-101, *99*,
186, 198, 223, 227-8

and life events 223-30

and marriage 224-6

and migration 48-9, 228-30

and negative feedback *see* negative
feedback in weight regulation

and night work 226-7, 252-3

and subconscious power of the
brain 27-8, 29, 77

in Vermont State Prison over-
eating experiment 16-17

weight loss

and appetite following bariatric
surgery 79

through bariatric surgery *see*
bariatric surgery

biggest losers in the lab 64

and calorie restriction 19-23, *23*

and dieting ix-x, xii, xiii, 19-20,
29, 61-7, 86, 214-17

difficulties in sustaining x,
19-30, 64-7

and DNP 75-6

and energy balance *see* energy
balance

through exercise 217-18

and infertility 94

and leptin 85-6, *86*

with lifestyle changes 217-19,
218

through lowering set-point
28-30, 102-3, 217-19,
218, 231-2 *see also* weight

set-point-based programme for
losing weight

and malabsorptive procedures
103, 104 *see also* gastric bypass

and metabolism/BMR 28-30, *29*,
64-5, *65*

and microbiome 230-31

and Minnesota Starvation
Experiment 21-2, 24

and negative feedback *see* negative
feedback in weight regulation

programme using set-point theory
see weight set-point-based
programme for losing weight

realistic expectations for 236-7,
276

shows on TV 63-4, *65*

surgery *see* bariatric surgery

and vegan diets 222

and vegetarian diets 222

weight set-point resetting
necessary for sustaining 28-30,
102-3

weight regulation

and calories *see* calories

of cattle 31-6, 60-61

through dieting *see* dieting

and insulin 197-8, 209

leptin as master regulator *see*
leptin

losing weight *see* weight loss;
weight set-point-based
programme for losing weight

and metabolic adaptation 18-19,
22-3, *23*, 26-30, *29*, 69-70, 77,
203, 248, 273

and Minnesota Starvation
Experiment 21-2, 24

and negative feedback *see* negative
feedback in weight regulation